Praise for *American Exodus*

Holding this book should feel like the touch of a cattle prod. But most of us have hides too thick to feel the shock and we will need several more, of ever-higher voltage, before we heed its message. For those with thinner skins, read it and be prepared.

— Clive Hamilton, author, *Earthmasters: The Dawn of the Age of Climate Engineering*

Standing apart from often-tedious tomes on climate change, Giles Slade's fast-moving *American Exodus* shows how it could create huge population movements that would bring US and Mexican catastrophe, perhaps amid Canadian opportunity. He ranges compellingly over North America and the globe, time and space, migration and settlement, plus history, science and art. He opens out thought by offering imagination, wit, detail, compassion and style. This book looks to become an accessible classic that might persuade you to move north before the rush. Watch Canada. And maybe watch out, Canada, too.

— Deborah Popper and Frank Popper, originators of the Buffalo Commons idea

American Exodus is the more polite title for a book that might have been called, as a chapter sub-head has it, "The Awful Truth" about how climate change will remake the settlement of a continent. Characteristically widely and deeply researched, Slade's new book argues persuasively that over the coming century much of the southern half and coastal zones of North America will become uninhabitable. The "exodus" of the title will then come to Canada— relatively green, wet, and mild. But as Slade points out, not all of Canada will be hospitable under climate change either. An engagingly eclectic meditation on the century to come, and the dramatic changes for which our countries and their leaders are woefully unprepared.

— Chris Wood, journalist and author, *Down the Drain: How We Are Failing To Protect Our Water Resources*

Giles Slade has never shrunk from talking about the next big idea. Now he takes on climate change and its potentially disastrous consequences for the United States. These include economic collapse and the turning of millions of Americans into postmodern "Okies" trying to cross their northern border into an unwelcoming nation. Slade makes plain that although America may still be the most powerful country in the world, in the face of recent natural disasters associated with climate change— urban heat waves, droughts and superstorms—the nation has looked like a hundred-pound weakling. As he paints a picture of a nation dangerously unprepared to face the current crisis, even fans of Slade's previous work may find *American Exodus* a very inconvenient truth.

— Edward Kohn, author, *Hot Time in the Old Town: The Great Heat Wave of 1896*

American Exodus assumes, unlike some, that global warming and climate changes are real threats to human survival. Instead of offering the usual standard response to this global issue, Giles Slade goes beyond this, and suggests provocative actions we need for the reproduction of life in North America. Slade shows that migration has always been an outcome of climate changes since the early 20th century, and projects that future American migration to Canada for survival is a likely scenario based on his rich analysis of the human migratory history of North America.

— Sing C. Chew, Helmholtz Centre for Environmental Research—UFZ and editor, *Nature + Culture*

AMERICAN
EXODUS

For Betty Slade…
still watching the world change after 94 years.
Enjoy this, Mom. Love and thanks.

AMERICAN EXODUS

CLIMATE CHANGE AND THE
COMING FLIGHT FOR SURVIVAL

GILES SLADE

new society
PUBLISHERS

Cover design by Diane McIntosh. Cover image: © iStock

Printed in Canada. First printing September 2013.

New Society Publishers acknowledges the financial support of
the Government of Canada through the Canada Book Fund (CBF)
for our publishing activities.

Paperback ISBN: 978-0-86571-749-7   eISBN: 978-1-55092-548-7

Inquiries regarding requests to reprint all or part of *American Exodus*
should be addressed to New Society Publishers at the address below.

To order directly from the publishers, please call toll-free (North America)
1-800-567-6772, or order online at www.newsociety.com

Any other inquiries can be directed by mail to:

New Society Publishers
P.O. Box 189, Gabriola Island, BC V0R 1X0, Canada
(250) 247-9737

LIBRARY AND ARCHIVES CANADA CATALOGUING IN PUBLICATION

Slade, Giles, author
American exodus : climate change and the coming flight for
survival / Giles Slade.

Includes bibliographical references and index.
ISBN 978-0-86571-749-7 (pbk.)

1. Climatic changes—United States. 2. Climatic changes—Social
aspects—United States. 3. Environmental refugees—United States—
Forecasting. 4. Migration, Internal—Environmental aspects—United
States. 5. Emigration and immigration—Environmental aspects—United
States. 6. United States—Environmental conditions. I. Title.

QC903.2.U6S53 2013   304.80973   C2013-904546-5

New Society Publishers' mission is to publish books that contribute in fundamental ways
to building an ecologically sustainable and just society, and to do so with the least possible
impact on the environment, in a manner that models this vision. We are committed to doing
this not just through education, but through action. The interior pages of our bound books
are printed on Forest Stewardship Council®-registered acid-free paper that is **100% post-
consumer recycled** (100% old growth forest-free), processed chlorine free, and printed with
vegetable-based, low-VOC inks, with covers produced using FSC®-registered stock. New
Society also works to reduce its carbon footprint, and purchases carbon offsets based on
an annual audit to ensure a carbon neutral footprint. For further information, or to browse
our full list of books and purchase securely, visit our website at: www.newsociety.com

MIX
Paper from
responsible sources
FSC® C016245

Contents

Introduction: Cooler Climes and Higher Ground xi

1. America's First Refugees 1

2. Transhumance: Mexico and California 31

3. Runnin' Dry 61

4. Seaboard Diasporas 87

5. Urban Heat 113

6. Drought in the Carbon Summer 139

7. Wind and Water 165

8. The Safest Place to Go 189

Conclusion: Cooler Climes and Higher Ground 219

Notes 231

Index 263

About the Author 271

Cooler Climes and Higher Ground

We are vanishing from the earth,
yet I cannot think we are useless or else Usen
[the "Great Giver"] would not have created us.
He created all tribes of men and certainly
had a righteous purpose in creating each.

GERONIMO (1906)[1]

Hey gringo, gringa! Wake up.

Climate change has already made a desert of northern Mexico. And droughts in Baja California Sur, Chihuahua, and Coahuila are expected to lengthen and intensify. Twenty million Mexicans currently live in conditions of acute food insecurity. The hardships in these border regions have launched massive migration into the United States since 1982.[2] Things in Mexico can only get worse. A recent Whitehall security report made this prediction about the future of the region:

> Increasing irregularities in the rainy season brought about by climate change will impact the groundwater level and have a disruptive effect on food production.... The disruption of cropland can result in undernourishment of the population, which increases susceptibility to infection, encourages displacement and ultimately could result in permanent migration.[3]

The aridity of Mexico's northern deserts has become so inhospitable to life that even the emblematic mesquite lizard has become a candidate for extinction; its fate is only a minuscule part of the same mass extinction event that now threatens all known life.[4] If you've visited Mexico more than once in the past 20 years, you've probably seen the change. The tropics are creeping irretrievably toward the poles as Texas and much of the Southwest also turns to desert.

Migration is the most common response of human populations to severe reductions in habitability, and the United States has never been immune to climate migrations. In the last century, several climate-induced migrations occurred across the Great Plains, and previously populated US centers have already returned to wilderness. (Strangely, collecting and visiting these ghost towns has become a leisure travel activity.) A similar pattern of outmigration is just beginning in California which now posts "deficit migration" statistics of about half a million people per year. Similar declines will soon encompass increasingly dry states like Texas. Metroplex areas like Dallas-Fort Worth, which have experienced decades of astronomical immigration, now have millions more people than can be supported by the radically declining water supply or the state's broken agricultural economy. Texas's current population far exceeds the state's carrying capacity. For Texans, water is now a more critical economic factor than oil.

In addition to drought, outmigration follows the destruction of coastal cities, as it did in New Orleans in 2005. More recently, Superstorm Sandy foreshadows similarly destructive events in the population centers of America's Eastern Seaboard. North Atlantic hurricane activity is increasing. Three once-in-a-century storms submerged portions of the Eastern Seaboard in just the first three years of the second decade of the 21st century. Much larger storms than Sandy are predicted. No one knows how much time we really have.

Still more troubling is the same combination of economic collapse and climate change caused by global warming that drove massive migration from Mexico into the United States after 1980. This single migration has already changed the demographic composition of both Mexico and the United States. Population growth projections assume a predictable constancy described by the word "stationarity," which names a kind of statistical status quo. The regularity of the statistical universe is then used to plot curves and make predictions from them. For example, projections of steady growth predict that by 2050, one in three Americans will belong to the loose-fitting racial cluster called "Hispanic" or "Latino."

When less than 50% of the United States belongs to the descendants of Europeans, the myth of "la Reconquista" will become a palpable reality. As the baby boom generation disappears, it will leave many fewer descendants. Fewer people will celebrate Columbus Day, and, in all likelihood, the lively and engaging festival of Día de los Muertos will replace the candy, costumes and consumerism of Hallowe'en. The official language of the United States may well change to Spanish, bringing the country in line with many more of its North and South American neighbors. It is easy to see that Mexican migration has already guaranteed further social change throughout North America.

Economic collapse always accompanies climate change, and it is this powerful combination that drives migration. In the United States in coming decades, both processes will continue their downward spiral until the habitability of America's most densely populated areas is substantially reduced. The carrying capacities of California, the Southwest, the Great Plains, the Northeast and the Midwest will decline radically. This is not b.s. An exact figure for the number of Americans vulnerable to future climate migration is impossible (for me) to calculate, but—ultimately—such an exact number doesn't really matter. Many-upon-many people will find themselves without the necessities of life: food, water, power, security for their children, and hope for the future. In desperation, they will realize their vulnerability, and then they will begin to move very quickly. It will begin as a trickle, but it will quickly become a flood. Of this group, a significant portion will move many times. They will go wherever they believe better lives are possible. At first, what constitutes a better life may be unclear to them. Gradually, however, their criteria will be reduced to a single, simple imperative: survival.

American Exodus claims that the movement of Mexicans since 1982 is simply the first stage of a climate migration that will eventually force Americans to leave their homes in the South and along America's coasts for the best climate refuge the continent offers: Canada.

There is probably very little that those in Canada and the United States can now do to prevent this process, but we can prepare for it, especially on a personal level. Collectively, although I believe the time to mitigate or delay humanity's suffering is past, we may still be able to limit its scope. To do this, we need a global mobilization on an order of magnitude never before imagined or attempted. As of yesterday, we need to dedicate ourselves—as an entire species—to the twin tasks of withdrawing atmospheric carbon and cooling our planetary home. Unfortunately, there is insufficient unified political will to accept this absolute necessity. My deepest fear is that it will require a major cataclysm—something much more powerful and destructive than Superstorm Sandy—before we accept what needs to be done. By that time, it could easily be too late.

In the meantime, it is very, very, very likely that a second stage of the ongoing Great Migration will occur. Certainly, climate change and economic collapse will drive outmigration from the continental United States into Canada. People will flee for their lives, just as 100,000 African Americans fled the south when the boll weevil changed Dixieland's cotton-economy. Just as with Mexican migration to the United States, the number of people in motion will be so large it will be impossible to stem the tide. America's northern border (along the 49th parallel) marks the longest unprotected international boundary in the world. Most of these 5,525 miles are unmarked and unpatrolled.

The legal framework that the United States have adopted to address climate disasters is a coal-burning antique "designed to repair and replace damaged infrastructure in a community's original location."[5] Already, Alaskan towns like Kivalina, Koyukuk, Newtok, and Shishmaref highlight the inadequacies of such recovery laws and strategies. For the natives of these communities—as for many others—there will be no recovery in their place of origin. They must simply move out of the way of the rising sea. Soon, many more Americans will also have to move. Since it is very easy to come north from the continental United States, Canadians can reasonably expect company. A lot of it.

They should now choose how best to greet their neighbors when the cataclysm brings them to the doorstep of "the true north, strong and free."

CHAPTER 1

America's First Refugees

*The…economic culture…had turned a continent into wealth,
had created vast fortunes, had made American agriculture
more of a business than a way of life, had taken immense
chances with fragile environments and had left many bills
to be paid by the next generation.*

DONALD WORSTER, THE WEALTH OF NATURE (1993)

(North) America the Beautiful

When I was a boy, my father got me to remember the names of
countries with little stories. He told me that North America was a
continental sandwich that had Mexico on the bottom and Canada
on the top. Confused about the role of the United States, I asked
him "What is it?" He said, "Tasty, kid, really goddamn tasty."

Today, from Tobasco, Mexico to Tuktoyuktuk on the Arctic
Ocean, our multiplicity of regionalisms, nationalisms and citizen-
ships have become irrelevant. In the face of ongoing, radical climate change, what is now most important is the sandwich we all
share. The world is changing. Although global warming has long
been an accepted fact in Africa, Australia, Canada, Europe, Mexico
and Oceania, in 2009 when I first planned this book, only 30% of
US citizens believed climate change was happening. In the United
States, I could not find a publisher for my work, so over the next
few years, I moved on to the task of researching, writing and selling

The Big Disconnect, a book about how the technologies invented and retailed by advanced capitalism has resulted in the worst social isolation that has ever existed.

By the time I finished that book, it was the summer of 2012. Suddenly, 70% of Americans had come to accept the fact that our planet is being altered by the human activities that produce greenhouse gases. This is really a simple idea, but it is as conceptually powerful as the belief in evolution, and the fact that it is now mainstream may have some fortunate consequences. Clearly, by 2012 the intensification with which greenhouse gases impact the meteorological exchanges in earth's atmosphere had become overwhelmingly obvious to most people. And no wonder, since these events are now happening more frequently and in very threatening ways. In North America, 2012 brought with it one natural disaster after another. By mid-July, I had two publishers on the line, both eager to re-examine my proposal for a little volume about the complex relationships between climate and human migrations across our home continent.

Climate change, of course, concerns us all. Most importantly, it concerns our children—white, black, brown, red, or yellow. Wherever they are, whatever language they speak, it is their lives and well-being that is at stake, and we must act for them, in their interest. To use terms previously reserved for a coming ice age or the threat of nuclear destruction: what we do now will determine if mankind has a future. We have reached the point where we must decide between being and nothingness and between apathy and action.

Climate Refuge/Climate Refugees

North America was originally populated by waves of Neolithic migrants who survived the last ice age by sheltering in small areas we now call "climate refuges." Having expanded throughout Asia as the world warmed, these groups crossed into North America as the ice receded. But the *modern* story of American climate migrations began early in the 20th century—with an environmental disaster of unprecedented proportions. At the time, it wasn't recognized as an environmental disaster. Even today, some argue that the word "disaster" doesn't really describe the event. Whether it does or it doesn't, in 1941, the storms ceased, slipping away like Camus's plague to wait patiently across the span of years for the next opportunity to wreak havoc on an unprepared populace. This time, it would wait 71 years—until October 2012, when large-scale dusters began returning to Oklahoma, and people like Associate State Climatologist Gary McManus began to use the word "disaster" once again:

> We have the extremely strong winds...[and] the timing is bad because a lot of those farm fields are bare. The soil is so dry, it's like powder. Basically...a whole bunch of topsoil [is] waiting for the wind to blow it away. It's no different from the disaster of the 1930s.[1]

The Galveston Flood of 1900 is often called a disaster, as is the San Francisco 1906 'quake and subsequent fire. Sudden and horrendous, both occurrences were cataclysms marked by rapid loss of life and massive damage to property, and both were in locations that could be pinpointed on a map. They were clearly disasters. What word, however, should be used to describe an ongoing scourge that spanned a dozen states but saw immediate fatalities numbered in the sparse handfuls? It was a very strange disaster. It claimed the majority of its victims in unexplained and offhand ways, like epidemics of measles that blew in on the wind or general wear on people's respiratory systems that would kill them via pneumonia or silicosis months or years after the dusters passed out of memory. In fact, for many, immediate death seemed preferable to being left struggling and alive in a degrading subsistence until years of malnutrition took the weakest and the hindmost. Suicides and mercy killings increased in the 1930s. Fathers, and even daughters, were known to take the lives of family members under the nerve-shattering stress of this relentless American calamity. To date, there has been nothing else like it in the recorded memory of our continent. It's possible that the Anasazi suffered in the same way before resorting to systematic cannibalism and then abandoning their elegant homes in the Southwest 200 years before Columbus. And geologists now also speculate that over 3,000 years ago, the Mayan civilization of the Yucatan peninsula came to an abrupt—and surely, anguished—end consequent to droughts accelerated by climate change.[2]

In 1981, 40 years after the last *black blizzard*, an historian—who was born in California just months after his parents had fled their ravaged Midwestern homestead—described what he had since learned about the event. It was, he wrote, the

> most severe environmental catastrophe in the entire history of the white man on the continent. In no other instance was there greater or more sustained damage to the American land…and its inhabitants.[3]

The same dry wind that had destroyed the mysterious Anasazi culture of the Four Corners blew back into the Great Plains in the 1930s, stripped it bare once again, and scattered its residents like the topsoil in which their livelihoods had taken root. When this happened, the Depression was already in full swing. In March of 1933, after two full years of drought, the sky turned black with the windborne dirt of "rollers" for all but nine days. In April, weather stations across the Plains recorded 179 of these rolling storms of dust.

It was as though a plague had visited the country in the worst of times. Eerie night skies bolstered the popular belief that a biblical punishment was being meted out against the Plainsman, who was experiencing cruel physical suffering on top of years of financial hardship and deprivation. At the time, the nation responded with a grudging, meager charity, condemning the victims of the calamity in the same way Eliphaz, Bildad and Zophar once condemned their neighbor, Job. There is a bitter significance to acknowledge here: long before Hurricane Katrina forced America to recognize how national emergencies are tainted by racial politics, an impoverished America only begrudgingly acknowledged its kinship and shared its bread with the native-born white Christian farmers who once supplied the entire nation with its sorghum, corn, wheat and cotton.

It is this lack of generosity—this first American example of *lifeboat ethics*—that delivers a powerful lesson for future migrants, whether you call them internally displaced people (IDPs), environmental refugees, or some other masking euphemism. In all cases, such regional protectionism is a demonstrably shortsighted, but nonetheless typical response that has deep implications for future generations because the decades to come will see the most profound climate change in the history of our species.

Dust Bowl Days

What was the Dust Bowl? What caused it? How did it begin?

Unlike climate change, the Dust Bowl had no Al Gore to issue stern and patient warnings about the near future. Not a single researcher turned his attention to the central problem of wind erosion until 1933, when an obscure ecologist named Aldo Leopold kickstarted the practice of conservation with his stern observation that civilization could not depend upon the "enslavement of a constant and stable earth," but relied rather on

> mutual independent cooperation between human animals, other animals, plants and the soils which may be disrupted at any moment by the failure of any of them.[4]

Leopold knew that long before man had arrived, the "southwesters" had blown across the Plains. When these "dusters" pass low over uncultivated patches of scrub or pastureland, fine grains of soil can easily become airborne and travel a thousand miles before settling. Leopold also knew that this process is greatly exaggerated when drought sucks all moisture from the earth, killing vegetation such as buffalo grass, whose stalks act as a windbreak and whose runners hold down the fine particles of clay and quartz.

My colleagues in Saudi Arabia call this kind of storm a *haboob* (هبوب), a "big blow," or a "blasting," based on the Sudanese regional verb "habb," meaning "to blow." I sat through a haboob once—in the front seat of a 4 × 4 in the Rub' al-Khali, which is 200 miles south of where Highway 75 ends with a terse bump in the northernmost reaches of the Empty Quarter. Here, the desert winds transport particles of silicate sand across southern Saudi Arabia and the United Arab Emirates before dumping it on Gujarat state in India or on the western deserts of Iran.

We had stopped, facing north along a north-south axis across the oncoming westerly wind, because my language tutor, an amusing Yemeni prince who usually taught European languages, wanted me to experience what he called "the Kingdom's lullaby," the side-to-side motion of a storm from the deep desert rocking our vehicle on its struts and springs. As the sun sank, we drank tea and smoked shisha with the windows open, watching a brown-orange cloud gradually obscure the entire horizon. We had plenty of time to extinguish the water pipe, and close the windows. The storm hit us full blast within the hour. At first, it lifted our truck precariously off its right-side wheels, suspending it for a few timeless moments at what must certainly have been its tipping point. "Alluha akbar!" my friend shouted when we dropped back—clunk—onto four solid wheels, "Why is God bigger?" I asked. It was a question that had puzzled me for a long time. "It really means God is Great," the Prince said. "He lifted us up. He put us back down. He is Greatest of all."

Even though it was now on four wheels, the truck rocked all night, and the roaring wind did not stop until an hour or so before dawn. When I awoke, the insides of my ears were numb, but I could see sunrise outside of my east-facing passenger-side window. On the driver's side, sand had collected to the level of the roof, making our Pathfinder (called a "Patrol" in the Kingdom) into a small, tan-colored hill. Shouldering the door open, I stepped into the sunlit, leeward crescent of an embryonic dune from which our truck was extracted only with great difficulty. Grinning like teenagers, we breakfasted under the azure Arabian sky, stuffing ourselves with tameese, foul, and Saudi champagne (mineral water mixed with apple juice) before unscrewing and extending two surplus Desert-

Storm entrenching tools to begin the long process of digging out our ride home.

"Mumtaz," my friend said when we were finally freed: "Wonderful." We drove silently northward as I ruminated about the accumulated raw force that created the Empty Quarter over millennia.

Although it is not as fine or as light as the white-grey, silicate sand of the deep desert, the *loess* soils ("loess" comes from a German word meaning "loose") of the Southwest nonetheless blow easily, especially in times of drought—and especially when the topsoil has been disturbed. By the 1930s, there was something new in the land, something that would have a profound effect on the topsoil.

Settlers had been coming to the Southwest since the 1820s. Historian James Malin has described how these dirt-farming pioneers experienced the same types of storms that plagued the Dust Bowl, though not as frequently. During a drought in 1860, for example, a wry Kansas City editor offered real estate for sale by the "bushel or acre...[since] necessity compels us to dispose of the fine bottom land now spread over our type and presses."[5]

White people were not the first to experience the hardships of flying dust. The Sioux called these storms *mak'op'o*, or earth (or dirt) clouds, and believed they signaled the presence of one of the most powerful Thunder Beings in their cosmology; the Sioux call him Wakinyan, the Storm-God, but the Cherokee call him Thunderbird (Asgaya Gigagei). The Cherokee held that only the lowest, "weather" layer of atmosphere in the Southwest was possessed by the Thunder Being. This demi-god was a powerful servant of the "Great Apportioner" who brought blessings and rain, but also did considerable harm to men if he was not placated in the Green Corn Ceremony every fall.[6]

In the years before the Dust Bowl, the Thunder Being had been generous. Long before the precise characteristics of the El Niño-Southern Oscillation were documented, their effects were obvious to those living through them. When the great drought of the 1890s ended, it was followed by a humid period lasting 30 years. By 1929, few people remembered the deprivations of the pioneers (or the

Thunder Beings themselves, who had lost a great deal through time and translation, becoming reduced to mere "dust devils").

Remarkably, until the early 1930s, if anyone turned their attention to erosion at all, it was usually to the type of erosion caused by water. Records turn up only one instance of a farmer who, during a period of minor drought, tried planting his cornfields at right angles to the wind to prevent his topsoil from blowing away.[7] But when the rain finally fell in Colby, Kansas in 1914, he abandoned the experiment. A German immigrant, Fred Hoeme, was probably among the first to understand the problem of wind erosion completely, but his insight didn't occur until 1933, and the chisel plow he devised to fight the problem wasn't widely available until late in 1935—by which time the rollers had already been coming for several years. The early 20th century was clearly a trial period during which agricultural techniques inappropriate to the Southwest came into wide use before being discarded. New methods, like the "dry mulching" once recommended by the Soil Conservation Service, were often adopted and then abandoned as farmers came to grips with the peculiarities of their thin soil that held moisture so poorly. Such misguided experimentation did a lot to ruin the region's topsoil.

And then, there was greed and technology.

Enslavement of the Earth

A new set of ecological values characteristic of the pursuit of agriculture as a corporate business developed in the increasingly capitalistic milieu of the early 20th century. Leopold succinctly described capitalism's exploitive ecological values as a naïve and temporary "enslavement of the earth." And historian Donald Worster, himself a child of the Dust Bowl, offers a set of three maxims that illuminate 20th-century man's alienation from nature:

1. *Nature must be seen as capital*…a set of economic assets that can become a source of profit or advantage, a means to make more wealth.
2. *Man has a right, even an obligation to use this capital for con-*

stant self-advancement.... The highest economic rewards go to those who can extract from nature all it can yield.... The social order should permit and encourage this continual increase of personal wealth.

3. *In pure capitalism, the self as an economic being is not only all-important, but autonomous and irresponsible.* The community exists to help individuals get ahead and to absorb the environmental costs.[8]

With this ideology in place, technology on the Great Plains affected the labor of domestic animals and human beings alike in bold new ways. Wheat harvesting machines replaced farmhands completely and turned the trickle of labor flowing out of the region into a flood that would last until well after World War II.

The "culture of technology" can be thought of as emerging from the ice ages—when it first began to facilitate speedy adaptation to climate changes. By the 1930s it had, of course, undergone several radical revolutions. Rail and steamboat transportation, for example, had proved especially decisive in the economic contest for dominance between the industrialized North and the agrarian South 60 years earlier. Then, with the introduction of Henry Ford's Model T in 1903, the union of capitalism and technology was complete. America remade itself into the world's technological and economic leader—a nation devoted to mechanization's promise of unceasing economic growth. It has remained so ever since.

Our love of gadgetry and machinery has its source in genetic adaptations that first appeared about 2.6 million years ago—the date of the oldest-known stone tools. Our technological adaptation has, of course, been reinforced by the succeeding centuries. Unique among species, we no longer have to wait for natural selection to accommodate us gradually to a new or changing environment. Instead, we adapt quickly, using the body of skills, tools and techniques we call "technology" and "culture." A recurrent goal of technological progress is to empower a single individual by allowing him to accomplish *alone* what would otherwise require the assistance of many people. In psychological terms, this individual empowerment acts as a powerful reinforcement of the individual human ego. At its most extreme, technology's tendency to empower the individual may be an intrinsic *dis*advantage because, by facilitating isolation, it encourages a decline in social capital.

How interesting, then, that those who practice technological adaptation professionally (engineers) control a much larger segment of the global economy than the legal and medical professions combined. The completeness of our faith in technological progress is unchallenged, probably because it has never failed us on a global scale (although writers like Joseph Tainter, Sing C. Chew and, most recently, Jared Diamond have shown that exploitive technologies have often been the cause of the collapse of *individual* civilizations). In the near future, our unquestioned optimism about the ability of tools, skills and knowledge to provide "technological fixes" for environmental problems may prove to be the great blind spot in mankind's reliance on this single, highly specialized adaptation. We may not be able to "geo-engineer" our way out of climate change.

In any case, the clock is now ticking.

The Scale of Technology's Impact

The Dust Bowl is a valuable example of how humans alter their environments. And it demonstrates a second intrinsic disadvantage of technology: its capacity for immense scale. Widespread adoption of a new technology can devastate an environment before enough people catch on and stop using it. By 1935, Paul Sears, an Oklahoma scientist, warned that the desert was expanding on the Great Plains, because farmers had

> reversed the slow work of nature that had been going on for millennia.... If man destroys the balance and equilibrium demanded by nature, he must take the consequences.[9]

H. H. Bennett, head of the Soil Conservation Service, went public in 1936, laying the blame for the Dust Bowl firmly at the feet of "power farming."[10] Bennett was deeply critical of how plows and tractors had destabilized the topsoil of the region. By this time, America was ready to believe him; weeks of winter storms had dumped chocolate brown snow on eastern cities (including Atlanta, where Margaret Mitchell changed the title of the novel she had just completed from *Tomorrow is Another Day* to *Gone With the Wind*[11]).

But the full implications of Bennett's claims were complex. Charles Angell had patented the "one-way disk plow" in 1923. Using this new device, a farmer could till much more acreage in a given amount of time. The Ohio Cultivator Company purchased Angell's patent in 1926, and, until Fred Home's chisel plow became available, the one-way plow was the standard device for farmers on the Plains.

Angell's invention made turning the already loose soil too easy. The first problem was that the plow's disk left "a pulverized, smooth soil" that was extremely vulnerable to wind.[12] But this was compounded by the fact that during the 1920s, Great Plains farmers developed the technique of summer fallowing—leaving their fields uncultivated for an entire season in order to restore the soil's nutrients. This resulted in better crop yields. But fallowing their fields didn't mean leaving them untouched. In the spring and fall, farmers turned the soil of the fallowed fields to eliminate weeds. This twice-yearly turning pulverized the soil again and again, removing groundcover that would have acted as a windbreak and roots that would have held the topsoil down.

Angell's one-way plow was usually purchased in a ten-foot length that taxed even a strong team of horses. But its invention coincided with the increasing popularity of tractors as an essential tool of agriculture on the Great Plains. By 1920, after only two years of tractor production, Henry Ford had sold 100,000 of his innovative Fordson units. Even the economic downturn of 1922 could not discourage Henry Ford—he responded first by making tractor purchases available on easy credit terms, and then by directly attacking the market share of his major competitors by slashing his Fordson's price from $625 to $395.[13]

When International Harvester's general manager, Alexander Legge, heard the news about Ford's price-cuts, he famously met the Fordson price and threw in a free plow to sweeten the deal for farmers. A three-way contest between Ford, IH and Caterpillar developed. (As the tractor price-war heated up, IH lost money on every tractor it sold.) As a result, tractor sales experienced explosive growth in the years before the Depression. (The success of American tractor sales resulting from cutthroat corporate marketing became the stuff of legend when "a natural born salesman," Alexander Botts—modeled, some say, on IH's Alexander Legge—appeared in William Hazlett Upson's *Earthworm Tractor* stories serialized by the *Saturday Evening Post* beginning in 1927.) During the Depression, tractor sales enjoyed unprecedented success. In Kansas alone there were just over 17,000 tractors in use in 1920; but by 1925 that figure had nearly doubled to 31,000; and by 1930 Kansas had 66,000 tractors. Despite the Depression, by 1935 the number of Kansas-owned tractors had risen to 71,000.[14]

Besides providing the farmer with a fast way to prepare his fields by himself, however, buying tractors put farmers deep into debt, as did buying the combine harvesters and trucks that farmers were also using. Because they made teams of horses and mules obsolete, power tools like the tractor freed former pasturelands for cultivation. In the years before the drought of the early 1930s, cultivation had expanded considerably; Angell's plow broke up the soil in many fields previously covered by grasses that made soil invulnerable to the wind.

In 1934, the storms that had become regular events a year earlier increased in size and intensity. With so much plowed-up soil, no wonder these intense storms spread western dust everywhere. When the winds blew again in 1935, they reached into the eastern United States, carpeting the coastal cities of New York, Boston, Washington, and Atlanta with fine silt before passing out to sea, where the brown-grey dust perplexed foreign sailors by settling on the decks of ships 300 miles offshore.[15]

During windless nights on the Plains, the Thunder Being himself now appeared high in the sky. Few people witnessing the strange nocturnal illumination realized they were seeing sunlight reflected by dust clouds beyond the horizon 50 miles above the earth in the mesosphere. The false dawn of the Dust Bowl was unlike anything ever before seen. It was very unlike the playful iridescence of the Northern Lights, which leaves the viewer with a sense of observing and participating in cosmic beauty. During the Dust Bowl, the eerie "noctilucence" of mesospheric clouds convinced many people that the end of the world was nigh; tent meetings and traveling preachers played to such widespread fears.

The enormous range and meteorological impact of all this blowing dust attests that "power farming" had delivered a revolutionary technology with a greater ability to influence the environment than any that had been available to farmers since the mastery of fire. Of course, at first there was no widespread understanding of how these new technologies impacted the earth and might have contributed to the Dust Bowl. And, at the same time, as a result of their sudden indebtedness, farmers on the Plains used their new tools as much as they could until, after years of drought, they might just as well have stood atop their barn roofs and thrown the precious resource of their farm's topsoil into the gathering winds.

There would be no respite for them until 1940.

Technological Unemployment

A further complication of the mechanization of farming that prepared the soil to blow were the changes affecting farm workers. In the early 1920s, before extensive mechanization, farmers harvested wheat using a combination of headers and header barges. A hired hand drove a team of five horses behind a device called the header, which cut the wheat. As the wheat fell, conveyor belts caught and lifted it onto the header barge where a team of "pitchers" stacked it using pitchforks. When the barge was full, the men offloaded the wheat onto the ground where it spent a month drying. Once the wheat had dried, a threshing crew fed it into a steam-powered thresher which separated wheat kernels from the straw.

These were seasonal and very labor-intensive processes. For well over a month, every farmer on the Plains needed a large, expensive crew to bring in his crop. His own operation was limited by the power of human labor to service an abundant crop, so until the Dust Bowl the Great Plains preserved the familiar two-tiered system of farming in which a lower class of tenant farmers provided a ready pool of seasonal labor for the independents. In a good year, this system paid enormous dividends. In 1926, for example, the wheat crop brought $10,000,000 to Texas County, Kansas alone.[16]

The promise of spectacular profits caused farmers across the Plains to expand their operations and mechanize quickly. Farmers who mechanized quickly experienced the powerful reinforcement of earning large profits from their crops with much less expenditure for personnel. The negative results, however, included increased indebtedness and an early form of widespread technological unemployment. (At the time, farmers' wives greeted their husbands' automation with unabashed rapture; the work of feeding teams of field-hands from dawn to dusk during harvest time was much harder and hotter than any agricultural job.) A similar wave of mechanization, expansion, and subsequent unemployment occurred after 1934, when federal agricultural programs put money into the hands of independent farmers. As one of them later said:

> I let 'em all go.... In '34 I had...four renters and I didn't make anything. I bought tractors on the money the government give me and got shet o' my renters. You'll find it everywhere all over the country thataway...the renters...got their choice, California or WPA.[17]

After widespread mechanization was made possible by government subsidies, the numbers of seasonal workers declined, so agricultural laborers began leaving the region in droves. Such technological unemployment was a popular theme of Depression labor unrest.[18] But the technological unemployment of Dust Bowl farm workers was not immediately obvious because the migrants took up similar work in other agricultural regions. In particular, they began to migrate to California where labor shortages had begun by 1931, when Mexican migrant workers were being coerced into volunteering for repatriation by rail.

As the variety of its crops demonstrated, California was a much richer agricultural zone than the Southwest, where wheat had been the main crop until 1930. After the widespread adoption of refrigeration cars around the turn of the century, markets for California produce began to expand across the United States; as a consequence, a sophisticated agribusiness infrastructure developed throughout the state. Then, in the 1930s, areas like the Central Valley expanded cultivation again, increasing the demand for experienced agricultural labor. Census figures show that about 24% of the 1,300,000 people who left the Southwest between 1910 and 1930 settled in California. In other words, even before the Dust Bowl and mechanization prodded them elsewhere, an extensive network of over 300,000 Southwestern relatives and friends were already on the West coast sending back word about life and work in the sunny state.[19]

The dam of Southwest outmigration didn't break until 1935, when the federal government, heeding the advice of researchers in the new field of scientific ecology, began buying farms and turning them back to grassland. These efforts coincided with the end of federal relief payments. As a result, many people who had struggled and barely held on for five grueling years decided at last to pack it in. It is remarkable that only one in four residents made the decision to leave. The others stayed out of dogged determination, out of shell shock, and out of resentment for all attempts to shift them from their hard-won family homes—a position best expressed by a contemporary farmer in New Mexico: "They'll have to take a shotgun to move us out of here. We're going to stay here just as long as we damn please."[20]

As early as 1933, Paul Taylor, co-author of a very timely and admirable book called *An American Exodus*, observed the replacement

of Mexican field-hands in the Central Valley by poor, white workers living in the crudest conditions. The massive growth in corporate "agribusiness" in California's Central and Imperial Valleys during the Depression was dependent on this ready supply of cheap,
migratory labor. The influx of experienced workers desperate for
jobs enabled California's "factories in the fields" to extract the maximum amount of profit from nature with the lowest possible investment.[21]

In 1933, Taylor described this new laborer as part of a class of
"drought refugees."[22] In the coming years, they flooded into rural
California in increasing numbers. Over 800,000 agricultural workers would leave the Southwest permanently in the next two decades, but during the years of the Dust Bowl and drought, many
more cycled through California eager to support their families and
find relief from the years of poverty that they had suffered on the
Plains.

Recently, historians have contested the refugee status of these
migrants, arguing that Taylor's "refugee image…exaggerates" because it

> elevates the tragic connotations of the migration, inviting
> inaccurate comparisons with…experiences of Europe after
> the two world wars or the even more devastating sagas of
> human misery that have so often plagued the Third World.[23]

But it is hard to agree with such revisionism. Moreover, laying the
choice of the word "refugee" at the feet of Paul Taylor is just plain
wrong. "Drought refugees," Taylor's description of the Plains residents he met in California's Imperial Valley, was a commonly used
phrase of the early 1930s. One of the first mentions of such refugees appears in a *Washington Post* article in 1931.[24] In describing the
first noticeable Okie migrants in California in 1933, Taylor simply
relied on a term that had already become common parlance because it aptly described the flight of Americans from their drought-
stricken homes even *before* the storms of dust began. Later, when
the dust blew, the press changed their wording, and for the next

decade a new phrase was applied across the board to all South-westerners who participated in the exodus from the Plains. Like the word "Okie," "Dust Bowl refugee" described anyone who left the Southwest during this period, even those who were simply fleeing drought and had never seen or experienced a dust storm.

Of course, *refugee* is a more loaded term than *migrant*, and the Depression was a much more plain-spoken era than our own—which uses the newly coined phrase "internally displaced person" to describe migrants leaving any region struck by disaster. In any case, if the accuracy of the word "refugee" is suspect, there can be no denying the scale of human suffering that took place on the Great Plains 80 years ago during the grueling drought that prompted a cluster of afflictions we still call the Dust Bowl. During the years of dust, families put their meager meals under their tablecloths at dinnertime. After saying grace, each family member shrouded his or her head under the cloth and ate dinner that way.[25]

During the blackest moments of the blackest year (1935), a Garden City Woman took the time to explain what living through a duster entailed:

> All we could do about it was sit in our dusty chairs, gaze at each other through the fog that filled the room and watch the fog settle slowly and silently, covering everything—including ourselves—in a thick, brownish grey blanket.... The doors and windows were tightly shut, yet those tiny particles seemed to seep through the very walls. It got into cupboards and clothes closets; our faces were as dirty as if we had rolled in the dirt; our hair was stiff and grey and we ground dirt between our teeth.[26]

Refugees?

The stress that residents of the Plains experienced during the thirties is undeniable as was their poverty, illness, malnutrition and suffering. But, should we call those who left "refugees"? The term is problematic in many ways.

Decades after the Dust Bowl, the United Nations decided that the word "refugee" meant a person crossing an international border to escape a well-founded fear of persecution on the grounds of religion, race, nationality, politics or social grouping.[27] Intuitively, this definition appears a bit overspecialized. Most people I speak with casually understand the word "refugee" as a simple description of someone in flight who is desperate for refuge. In the 1970s, renowned ecologist Lester R. Brown coined the phrase "environmental refugee" to describe desperate people who are *not* fleeing persecution but instead who are running from the destruction of their environment. Accepting this new category of refugee is really an acceptance that "in the past couple of decades...environmental degradation started to be included as a threat to human security."[28] Such refugees are expected to increase in the coming decades. Responsible researchers guesstimate that, worldwide, by 2050, there will be between 200 million and 250 million environmental refugees and about 1 billion homeless people among a world population of 9 billion.[29] Today, such numbers seem too large to describe the number of people driven from their homes. Mexico's population currently exceeds 100 million; America's exceeds 300 million. But by mid-century, it will be as though an entire country will have been torn from its roots and set in motion. Nonetheless, "environmental refugee" may itself be a deeply misleading term because, as Sing C. Chew has brilliantly demonstrated, even in cases where environmental collapse forces migration, it is inevitably preceded by economic collapse. In this way, "ecological limits" Chew writes:

> become...the limits of socioeconomic processes of empires, civilizations and nation-states, and the interplay between ecological limits *and* the dynamics of societal systems defines the historical tendencies and the explanatory trajectories of the human enterprise.[30]

Although "environmental refugee" clearly describes a growing number of people in flight from environmental calamity, the United Nations does not yet admit the existence of such migrants; as a result, recent figures are hard to come by. In 1997, it was estimated that there were 22 million traditional refugees and over 25 million environmental refugees. Two years later, the totals changed to 27 million environmental refugees and 25 traditional ones. If these figures are accurate, it is clear that the proportion of environmental refugees is increasing rapidly. Some expected the total number to double by 2010 and then to grow exponentially

> as increasing numbers of impoverished people press ever harder on over-loaded environments. Their numbers seem likely to grow still more rapidly if predictions of global warming are borne out...[then] as many as 200 million people [would] be put at risk of displacement.[31]

To my mind, the most significant thing about these figures is that they suggest future environmental refugees will not flee just Darfur or some other, distant location in the developing world. When global warming and its resulting climatic changes impact our children, North Americans will become as vulnerable as the residents of Africa. These stakes make it worth taking a careful look at why the Okies fled or did not flee during the Dust Bowl, as well as what the consequences of their choices turned out to be.

Reasons to Stay

Some people stayed on the Plains simply because they had too much to lose; some left because there was too little to hold them. Many farmers had everything they owned invested in their land. Leaving meant abandoning themselves and their families to the terror of an enduring poverty. In addition, these sons and daughters of the sod-busting pioneers had powerful community ties and a strong sense of responsibility. All these were admirably expressed by farmwife Caroline A. Henderson in *The Atlantic Monthly*:

> To leave voluntarily—to break all these closely-knit ties... seems like defaulting on our task. We may have to leave. We can't hold out indefinitely without some return from

the land, some source of income.... But I think I can never go willingly or without pain.... There are also practical considerations that...hold us here.... Our entire equipment is adapted to the type of farming suitable for this country and would have to be replaced at great expense.... I scarcely need to tell you there is no use in thinking of either renting or selling farm property here at present. It is just a place to stand on.... We could realize nothing...from all our years of struggle with which to make a fresh start.[32]

At first, those who left the Southwest during the initial drought were landless, agricultural workers—those least invested in the land. Among these, there were more men than women and most were under 30. A cohort of young, single men was the first to leave the Plains because of the meager quality of life available to them if they remained. Nothing but human affection prevented them from searching for a better life elsewhere. After this first group, young families began leaving to protect their children, who, as a group, are especially vulnerable to the diseases that follow upon the extended malnutrition characteristic of environmental disasters.[33]

Many migrants did not act alone, but participated in a "chain migration" typical of refugees throughout history. In desperation, the exo-dusters followed kinship or social connections in the trek west. Despite John Steinbeck's portrayal of the Joads, these Southwesterners did not usually bring the elderly with them, although they, along with the children, were the most vulnerable to the respiratory complications of flying dust. It was mainly young families who went West, and often several families traveled together in a mutually dependent cluster. They relied on advice from their predecessors about work, food, housing, and relief in the new land. Very often, like the immigrants who came to America and formed the Little Italys and Chinatowns of the Republic, they settled close together.[34]

It's useful to remember that this was a period well before the great leveling of American social capital described in Robert

Putnam's classic study *Bowling Alone*. In rural communities, people still had an assortment of close friends and neighbors on whom they relied completely in times of need. To refuse such help was to risk ostracism, and this, it should be remembered, is the fundamental utility of the social capital we have lost. Ironically, such capital is completely opposed to the emerging capitalist ideal of community in which "the self as an economic being is not only all-important but autonomous and irresponsible."[35] In times of emergency and stress, it is human connections—responsible relationships between members of a community—that make the difference between survival and being buried in the dust.

Don't Expect a Welcome Party

It was good that the Okies could rely on one another, because no one in California welcomed them. Although it had been hit less hard than other states following the Crash of '29, California responded to the employment shortages of the Depression by "repatriating" Mexican migrant workers. They did this to eliminate the costly "relief-harvest labor-relief cycle" that had developed among the Mexicans who remained in California in the periods between their seasonal jobs.[36] Beginning in February 1931, more than 150,000 Mexican citizens and their *pochos*—American-born children—were shipped to Mexico City by the Southern Pacific Railroad at a cost of $14.70 each.[37] Of course, just as in our own time, once the possibility of American jobs disappeared, many Mexicans were happy to leave the United States; they often traveled south voluntarily in a cavalcade of old cars similar to the ones that would soon bring the Okies *to* California.[38] This Mexican exodus preceded the Okie invasion, and it created a labor shortage that lasted until 1938. It was this vacuum of labor that the indigent Okies, desperate for work, would soon fill.

Significantly, the "repatriados" incident did something else. For Depression-era Californians, it stigmatized migrant workers, identifying them as undesirable aliens whose services were too expensive for a society attempting to recover from hard times. At its

beginning and again at its peak, the Okie migration met an entrenched set of what political scientist Garrett Hardin once called
"lifeboat ethics." Pressed by economic necessity, many Californians
urged conserving their state's resources and favoring native citizens
over outsiders who, they believed, would just consume the state's
generous relief resources. Even if these outsiders were Americans,
they were seen as "immigrants," and this negative image impacted
the Okies in several ways.

The strangest incident involved flamboyant LAPD Chief James
(Two Gun) Davis's "bum blockade." Davis was acting on recom-
mendations made in a 1935 report by the state relief agency's *Committee on Indigent Alien Transients* that reflected the popular fear
that California would become a "dumping ground" for the nation's
indigents. The report recommended the establishment of *Peace
Officer Stations* along the state's borders to prevent hobos, bindle-
stiffs, fruit tramps, and tin-can-tourists (those in tin lizzies or old
Model T cars) from taking up residence in California and becoming
a drain on the public purse.[39]

In February 1936, Chief Davis took it on himself to stem the tide
of indigent Okies by dispatching 136 police officers to the points
where the transcontinental highways entered California. California's Indigent Act of 1933 enabled authorities to refuse entry to
anyone with less than $100 in their pockets—and in 1933, very few
people walked around with $100. The law was rarely enforced until
1936, when Davis's newly formed "foreign legion" trawled newcomers at the main border points to find those traveling with "insufficient" funds. These unfortunates were roughed up, fingerprinted,
photographed and turned around in large numbers.

Ironically, Davis himself was a former Texan sharecropper. He
had earned the nickname "Two Gun" by winning both the right-
handed and left-handed pistol championships of the United States
in 1932.[40] Because he had successfully used a similar technique of
rousting criminals to rid L.A. of eastern mobsters, reaction to
Davis's innovative "blockade" was tolerant at first. Even the *New
York Times* viewed Chief Davis's efforts patiently, explaining that

Southern California...has suffered from the hitherto unchecked migration which has added a heavy relief burden
to the normal load...the financial loss through this cause
[is]...$5,000,000 a year.[41]

But soon, neighboring states like Arizona and Oregon began to object, fearing "a backwash of vagrants."[42] And 400 miles to the north,
residents of Reno, Nevada, became so annoyed by the LAPD's arrogance that they erected a large sign on their city's outskirts reading: "Stop! Los Angeles City Limits."[43]

Davis countered his numerous critics by claiming that

65 to 85% of the indigents entering California come eventually to Southern California [and] financial loss to this area
from this class of immigration is conservatively estimated at
$1,500,000; [and] to the state [at] $5,000,000. The hordes of
immigrants are not coming for work. They are coming with
the idea of getting on relief rolls, begging or stealing.... We
confidently expect a 20% decrease in the crime total of the
next 12 months.[44]

Strangely, no one but Chief Davis himself could find a source for
these figures. Concerned about the LAPD's expanding jurisdiction,
the California State Police sued California's second largest police
force. Then, with criticisms pouring in from journalists, legislators,
and civil liberties groups, the "bum blockade" ended on April 7th as
abruptly as it began.[45] Still, the image of Okies as shiftless, shirk-
ing "immigrants" eager to take a free ride on the generosity of their
more "American" hosts was left to percolate in the minds of many
Californians for a few more years until a strident antipathy to the
migrant influx began in earnest.

Uncharacteristic rains hit the San Joaquin Valley early in 1938
and lasted for weeks. The ensuing floods engulfed the Upper Valley
and washed away most of the Okie "ditch-bank camps." These were
haphazard shantytowns overlooking the fields where the migrants
obtained piece work. The rains left these families hungry, sick, cold,

wet, and poorer than ever before. A wave of coverage in the popular press generated charitable relief activities by the Simon J. Lubin Society and the Emergency Flood Relief Committee, but the flood's destruction of California crops was followed by a second Agricultural Adjustment Act, which reduced California's cotton-growing land from 618,000 to just under 350,000 acres.[46]

Suddenly there was an uncomfortable and problematic glut of agricultural workers throughout California. Many of them had been in California since 1935, so they qualified for the $40 per month state relief fund. This became a highly charged political football in California politics and had a pronounced impact on the already hard-pressed Californian taxpayer. Nonetheless, storms and unemployment left all the migrants needy. Despite state relief efforts, the costs of health care, shelter, food, and education stretched county budgets tissue-paper thin throughout the San Joaquin and Imperial Valleys. Local taxes in Fresno, Kings, Madera, Kern and Tulare Counties escalated sharply. Everyone blamed the Okies. During the peak of this resentment, the "immigrants" were renamed, and just as often were called "Okies"; the exo-dusters were also openly referred to as "the reliefers," "the chiselers," and "the freeloaders." In June, a group of Kern County businessmen met in Bakersfield and formed the *California Citizens Association* (CCA) to combat the migrant problem. Their petition to Congress began with these words:

> California is faced with economic chaos and financial ruin through the influx of thousands of families, displaced in other states and pauperized by the Depression.[47]

These businessmen demanded that California encourage the Okies to return to their home states; there, they would be eligible for *federal* relief funds. In the coming months a statewide movement to garner support for the CCA charter won the signatures of over 100,000 Californians (at a time when the state's population was barely 6 million). The rhetoric and tactics of the propaganda used against Okies ramped up during this period and migrants found themselves excluded from clubs and service organizations in many small Californian towns.[48] "Despite their whiteness," one historian wrote

> the Okies became identified in the minds of rural Califor-
> nians as field workers. Field workers had always been viewed
> as racial inferiors in the social order...[so]...the Okies in-
> herited the same racial prejudice...hitherto applied to...
> minority groups.[49]

The full extent of the Okie race-caste identification in California is powerfully illustrated by an incident in the summer of 1939: signs appeared in the lobbies of the San Joaquin Valley's segregated theaters reading "Negroes and Okies upstairs."[50] That same year in Tulare, Kings and Yuba Counties, officials once again deployed the Indigent Act (which had been revised in 1937) to rid themselves of "crooks coming into our county from the Dust Bowl."[51]

One reason the Okies were vilified, of course, was that as Cali-fornians were attempting to recover from the impact of Depression, they were eager to conserve local resources for Californians alone. This is a transparent expression of what since 1974 has been called "lifeboat ethics," and it began in California, not with the Okies, but with the exodus of Mexican workers.[52] It is important to remem-ber the "repatriados" incident because one of the first impulses of "lifeboat ethics" is to define membership in the "legitimate" lifeboat group and to exclude all those who are not entitled to the limited resources of the privileged few. In this connection, it should be re-membered that the racial purity requirement for citizenship de-manded by Germany's *National Socialists* began as a political tactic first to exclude and then seize the property of Germans of Jewish descent in the impoverished, Depression-era Germany. There is nothing novel or pretty here. This is what happens when people get afraid in hard times.

The second reason the Okies were vilified in California is more fundamental, even archetypal. Unfortunately, there is a recurrent human tendency to blame the victims of tragedy and misfortune. According to sociologist Michael Lerner, this is a kind of knee-jerk reaction that serves the purpose of shoring up our belief in a "just world" in which we are rewarded for our efforts by the certainty that completely undeserved misfortune will *not* befall us. The story of Job attests to the antiquity of our human tendency to accuse victims in order to support our belief that most suffering is deserved because "affliction cometh not forth from the dust."[53] Lerner explains it this way:

> The observer who sees a victim's fate as entirely deserved does not...experience any conflicts concerning the potentially costly consequences of intervention. There is no implicit threat to the image of one's self as a good citizen for failure to intervene, no sense of impotence at being unable to compensate the victim and punish the inflictor of injustice, no reason to risk one's safety or...resources to restore justice.[54]

Long before Lerner, however, John Steinbeck thoroughly understood the human forces at work during the Dust Bowl migration. And he succeeded in describing them with the enduring eloquence of a Nobel laureate:

> There was panic when the migrants multiplied on the highways. Men of property were terrified.... Men who had never been hungry saw the eyes of the hungry. Men who had never wanted anything...saw the flare of want in the eyes of the migrants.... Men of the towns of the soft suburban country... reassured themselves that they were good and the invaders bad, as a man must do before he fights. They said, those goddamned Okies are dirty and ignorant. They're degenerate, sexual maniacs. Those goddamned Okies are thieves. They'll steal anything. They've got no sense of property.[55]

Ironically, despite the bitter resentment that accompanied their move to California, the Southwestern migrants had a profound impact on political life in their new state. Their plight highlighted the exploitive practices of corporate farming and enabled the beginnings of California's agricultural strikes and boycotts that would culminate generations later in the labor victories of César Chávez.

More immediate results came from the Southwesterners' enduring loyalty to Roosevelt's Democrats. Beneath the financial considerations expressed by the California Citizens Association was a stark numerical political reality. Because they were not enfranchised voters, Mexican migrants had never threatened California's status quo. Mexican American migration did not overwhelm a majority of Americans until the issue of Mexican migration reached a nearly hysterical pitch in the anxiety, stress, and confusion following 9/11.

The bill proposed by Jim Sensenbrenner in the House of Representatives in 2006 was a panicked response to the rising tide of Mexican migration. The desperation of the period can be measured by Sensenbrenner's excessive proposals that illegal aliens should receive a prison sentence for remaining in the United States without documents and that legal American residents should be so imprisoned for *helping* any "indocumentado."

In California in the 1930s, the migrant Okies were all American citizens. This was their one great advantage. By sheer numbers, Okies so altered the state's voters' lists that in 1938 they forced the ouster of the pro-grower Republican administration of Frank Merriam. Democrat Culbert Olson was elected with a mandate of bringing Roosevelt's New Deal to California. Olson played very well among the migrants because he took a long view of their suffering, seeing them in a much different light from many in his state:

> They have fought despair...and...found themselves stricken with poverty. They have [come] to California to...get a foothold.... While they are being acclimatized and assimilated... we had to levy some taxes...because we don't want them to starve.... [But] they are good industrious American citizens like the rest of us...and...they believe that we should work and earn our living by the sweat of our brow.... It [just] takes a little time.[56]

Migration and Climate

In addition to the major impact the Okies had on regional politics, the significance of this migration also has profound implications for North Americans today. A few years back, a Washington think-tank concerned primarily with issues of national security was the first to report publicly that in the United States "by virtue of its large size and varied geography," the process of climate change has already begun. According to this report, *The Age of Consequences*, we are now experiencing

> a wide range of severe climate-related impacts, including droughts, heat waves, flash floods, and hurricanes.... The western United States, southern Europe and southern Australia will experience progressively more severe and persistent drought, heat waves and wildfires in future decades as a result of climate change.[57]

As these events increase in intensity, the report notes, they will increase in cost and draw down America's reserves. Each new crisis will reduce the economic resilience of state and federal governments and increase the country's vulnerability to climate change. Unfortunately, there is no relief in sight for more than a generation, since even the most radical reduction in carbon emissions now would have no noticeable results for about 30 years.

For those of us living today, the Okies and their ordeal during the Dust Bowl offers the lesson of how a large group of North Americans responded to the challenge of an environmental disaster caused by the ignorant use of a powerful technology: tractors. This monumental error on the Plains forced over a million residents to abandon their homes and migrate across the continent. Because of their numbers, this migration taxed the resources of every area in which they settled—California as well as the Northwest and the northern Midwest. In each area, the migrants impacted the politics and culture of their new homes. Their numbers, their desperation, their suffering, *and* the resistance they met are warning beacons to all of us when we consider North America's future in our own era of climate change and impending migration.

The Age of Consequences report presents three scenarios concerning life in America during the next 30 years. Following is the mildest version and the only one that downplays the consequences of internal migrations throughout the United States. Unfortunately, this is *by far* the least likely outcome, since (as the report itself notes) climate change is already occurring at a much greater rate than predicted:

> The United States…will not experience destabilizing levels of internal migration due to climate change, but it will be affected…. Tropical cyclones will…force the resettlement of people from coastal areas in the United States [with] significant economic and political consequences…. The United States will also experience border stress due to the severe effects of climate change…. Northern Mexico will be subject to severe water shortages, which will drive immigration into the United States in spite of the increasingly treacherous border terrain…. Damage caused by storms, and rising sea levels in the coastal areas of the Caribbean islands… will increase the flow of immigrants…and generate political tension…. In the developing world…the business and educated elite who have the means to emigrate…[will] do so in greater numbers than ever before.[58]

In other words, very soon a lot more people could begin clamoring to share the lifeboat called *El Norte*. Who we are and what we become will be determined in large part by how we respond to the demands of our neighbors in their time of need. This will be the first challenge.

The second challenge will come when national resources are depleted; richer and more fortunate American states will resist migrants from states where fires, desertification, drought and floods have made it impossible to sustain habitability for large populations.

The question that will then arise will be: "Will America continue?"

CHAPTER 2

Transhumance: Mexico and California

...like Indian Joads, they have fled the Mexican Dust Bowl.

RUBEN MARTINEZ, CROSSING OVER (2001)

Mexico is being bludgeoned by climate change. In coming decades, the US government anticipates increased illegal migration from Mexico as extreme weather continues to flood southern states like Tobasco (80% of which was covered in water in 2007). Similar floods destroyed the bean harvest in states on the Pacific Coast in 2010. In the north, the extreme temperatures responsible for killing crops in Texas and other southern drought states are now squeezing every drop of moisture out of Mexico's agricultural region. The best projections have it that an additional 2% of Mexicans aged 15 to 65 leave for the United States every time crop production goes down another 10%.[1]

Meanwhile, America is using a current lull in northward immigration to build better fences and to tighten security along the densely populated borderlands of California, New Mexico and Texas. In 2012, these efforts led Department of Homeland Security (DHS) Secretary Janet Napolitano to reassure Americans that the border with Mexico is "more secure than ever." But it's not the impenetrability of America's borders that's responsible for the current downturn in Mexican immigration. To speak frankly, it's patently

ridiculous to believe that any fence can keep desperate and determined Mexican and Central American migrants out of the United
States.

The current downturn has an obvious economic cause. Jobs became very scarce in America following the financial crisis of 2007.
Consequently, by 2012, illegal immigration from Mexico fell to zero
(possibly even less than zero). In 2007, the Pew Hispanic Center
reported that 12.6 million Mexico-born individuals lived in the
United States (this figure does not include American-born children). By 2011, that number had declined by 600,000. The chance
to make money in *el Norte* and then send it home to loved ones in
Mexico was the major "pull factor" driving legal and illegal migration prior to 2007. It's significant that when the economy picked up
briefly in 2009–2010, illegal immigrants once again began hiking
north through the most unprotected parts of the border, along the
eastern portion of the 300-mile beltway Arizona shares with the
state of Sonora.

In other states, America's 1,969-mile border with Mexico was
and is quite difficult to penetrate, so Arizona remained (until
2010) the only state in which illegal crossings continued to rise.
The shooting death of Robert Krentz after a confrontation with
Mexicans crossing through his border-country ranch crystallized
the anger Arizonans were feeling about illegal immigration. It resulted in a new state law—SB 1070—that allowed police in Arizona
to question "suspicious" people about their citizenship and immigration status. But Arizona's anger was not restricted to framing a
new law. The Minutemen, a group of armed vigilantes whose leaders are associated with America's neo-Nazi movement, stepped up
their patrols of the desert southwest of Tucson, and they are still
active as of this book's publication date.

Together, the Minutemen and the state law clearly indicate that
a level of genuine desperation and hysteria had been reached in
America's response to illegal immigration from Mexico. Although
many unemployed and desperate Mexicans had previously used
the United States as a revolving door for working sojourns, Mexi-

cans began coming to America and *staying* in large numbers after an economic event called "la crisis" gripped their own country in the early 1980s—around the time that Mexican translator, poet and journalist Ilan Stavans wrote: "Elsewhere—that's where I belong: among the vast diaspora."[2]

Mexican migrants are not usually considered "environmental refugees," and their movement does not popularly constitute a "climate migration." But it is an historical truism that significant climate change usually precedes a financial downturn—and this combination of factors causes people to leave their region. The crushing expense of repairing infrastructure tips the economic scales of an area into financial ruin. The massive migrations consequent to such ruin thus have *joint* causes: financial chaos *and* ecological devastation. Poet José Emilio Pacheco describes the environmental destruction that has afflicted his country for most of the past century:

> When there is not one tree left,
> When everything is asphalt or asphyxiation
> Or badlands, stony lifeless ground
> This will once again be the capital of death.[3]

As we'll see, the same combination of financial collapse and environmental devastation that is disassembling modern Mexico in front of our eyes has also initiated every known example of "failed state status," and is, indeed, causing the migrations that history has taught us to expect. Environmental collapse follows economic collapse much like the aggrieved drunk who brings a gun to the last stages of a wild party. No one can agree on when exactly things began to turn sour, but no good can possibly come of it. Anyone who is still sober leaves. In the same way, failure of the state almost always spurs mass migration.

Failure in Mexico

A father-and-son team who study Latin migration to America agree on the fundamental premise that underlies the complexities of contemporary Latino migration to the United States:

The migration of Latin Americans, though a highly visible, politicized, and controversial phenomenon, is…very poorly understood by the US public, even by most policy makers.[4]

To clarify the entangled push and pull factors that continue to bring Latinos to America in unprecedented numbers, some backstory is essential. This is all very recent Mexican history, but, because of the language barrier, it is largely unfamiliar to non-Latinos.

In the early 1980s, Mexico experienced the first round of a prolonged tailspin of economic woes that has not yet bottomed out or ended. The crushing poverty that resulted from these disastrous events is what first exposed Mexico to the stupid, savage rule of the modern drug cartels. In an era when car-bombs are deployed on the streets of Ciudad Juárez, ordinary Mexicans feel their country's institutions, lives, and infrastructure threatened daily by the drug cartels. In addition to a pervasive drought and a diseased economy, Mexico has become a place where the criminal behavior of the drug cartels has become, as one-time Mexican President Felipe Calderón put it, "an activity that not only defies the state but seeks to replace the state."[5] For the cartels, "business is no longer just the traffic of drugs," Calderón warns. "Their business is [now] to dominate everyone else."[6] Mexico's chronic political and social instability began when the country first plumbed the depths of poverty with the abrupt puncture of its inflated economy in 1982. There had been early warnings of collapse as early as 1976—when the peso was devalued—but at the time, the country anticipated tremendous revenue from seemingly endless offshore oil deposits. *El boom petrolero* (1976–1982) was an unjustified period of infrastructure megaprojects, widespread importation of foreign goods, and expensive government subsidy programs. One government program supplied fertilizers and pesticides to farmers; it was aimed at guaranteeing basic nutrition to all Mexicans, but ended up having disastrous, unforeseen effects (more on this, shortly). Using borrowed money, the oil boom financed a half-dozen years of reckless prosperity and payola. Then, like a gambler whose luck has suddenly changed, Mexico was left with an enormous tab.

Researchers for the *Mexican Migration Project* pinpoint the precise moment when Mexico's bubble burst:

> The Arab-led oil cartel disintegrated and world petroleum supplies rose causing the price of oil to drop precipitously.... A financial crisis in the summer of 1982...forced a devaluation of the peso and unleashed successive rounds of hyperinflation that persisted through the 1980s which came to be called the lost decade (*la década perdida*).[7]

The remainder of the 1980s disappeared into a coma of anxiety and hunger. By 1989, though, Mexico was struggling to its feet in a determined effort to inspire investor confidence. But the Chiapas Revolution in 1994 ended the country's brightening economic prospects by frightening all foreign investment out of Mexico. By 9/11, the crisis that began as the "lost decade" had lasted 20 years, with only a short recess. From the beginning, this crisis completely dissolved any elements of economic security for ordinary Mexicans. In February 1982, personal savings disappeared into an abyss of devaluation when *overnight* the peso lost 30% of its purchasing power. Mexicans with any remaining financial resources quickly deposited them outside the country.

As the banks collapsed or closed, inflation soared. Food prices, especially those of tortillas and bread, doubled and then doubled again. Manufacturing and retail sales faltered when many small businessmen, unable to pay foreign dollars for their imported inventories or materials, went bankrupt. Across the nation, jobs evaporated, and 1,000,000 Mexicans suddenly found themselves out of work. As a result, the national index of poverty *doubled*.

Charting the history of the Mexican crisis as it happened, political scientist Judith Hellman described how, with devaluation, the country's foreign debt skyrocketed: "Mexico owed 234 billion pesos at 1981 rates, but 405 billion at the 1982 rate of exchange."[8] In the early spring of 1982, Pemex (Petróleos Mexicanos, Mexico's nationalized petroleum company) determined that oil revenues for 1982 would amount to less than half the amount projected in the national budget. Already very poor, Mexico became suddenly destitute, which caused infrastructure projects and social subsidies throughout the country to end abruptly. Simple food security became an everyday issue for ordinary citizens. The social impacts were so complete that the crisis was personified as a tangible villain demonizing the lives of ordinary Mexicans. Reuben Williams, an award-winning journalist and a child of Mexican migrants, recalls:

> The darkness lingered on. Assassinations, corruption, street crime. In everyday conversation, Mexicans referred to the phenomenon as, simply, "la crisis".... People said that la crisis was responsible for every malady that afflicted them, a *deus ex machina* that was cause, not effect.... Because of la crisis you borrowed a thousand dollars and risked your life sneaking across the U.S.-Mexico border.[9]

The "Green Revolution"

Mexico's economic collapse paralleled the nation's simultaneous environmental devastation. In the countryside, the Green Revolution that transformed Mexican agriculture in the 1960s began revealing unforeseen consequences in ordinary small-holding farms by the 1970s. Before the oil boom, peasants were given fertilizers, pesticides and special seed stocks that made their land magically produce many tons more corn, beans, wheat, sorghum, soybeans and cotton. Using similar techniques, the Green Revolution made Mexican agribusiness possible because it created an agricultural surplus in the semi-arid northern and northeastern zones of Mexico. But these large-scale chemical inputs were accompanied by huge water management projects that transformed the national model of subsistence agriculture.

Water has always been a powerful instrument of social change in Mexico; with the Green Revolution, water transformed the "irrigation districts" of the north into centers of modern export business that grew major commercial crops including sugar cane, vegetables, alfalfa, soybeans, cotton and wheat. Unfortunately, most of Mexico's farmers were not large corporate farmers: "From the 1960s onward, of the 6.2 million hectares of Mexico's irrigated farmlands, [only] 2.8 million were irrigated by 27,000 small producers with private or communal water and land."[10] Most Mexican farmers of the day were *campesinos* or *ejidatarios*, subsistence farmers who grew personal crops on tiny privately or cooperatively owned plots of land. As payment of farmers' accumulated debts, much of this land was eventually acquired by Pemex, Mexico's banks, or international agricultural combines. Peasant farmers who began the Green Revolution in Mexico never got to enjoy the benefits of Mexico's irrigation megaprojects. In fact, with the exception of ancient, Moorish low-tech hydrological technologies (*qanats* and *shadufs*), Mexican peasants weren't really involved in irrigated agriculture at all:

> They plant[ed] during the rainy season and stretch[ed] their meager water resources over the dry months. The government's core agricultural and irrigation investments…passed these producers by, leaving them vulnerable to climatic fluctuations and to prices…set in…commodity markets.[11]

Moreover, although the Green Revolution radically increased Mexico's food production, not all regions became self-sufficient. In some areas, the situation was desperate long before the economic crisis broke. During the 1960s, despite billions spent developing an indigenous agribusiness, hunger was widespread in the South, where "83 percent of all the farmers…could maintain their families only at a subsistence…level."[12] For this reason, an excellent national system of food subsidies was instituted in the 1960s. Perhaps the best social program ever created in Mexico, it guaranteed basic nutrition to all Mexicans. But its abandonment in the early days of *la crisis*, left the poorest of Mexico's 90 million citizens without food to feed themselves or their children. So after 1982, they began to move north in great numbers, pushed from their homes by hunger and despair and drawn to America's affluence which was so great their American-born children would become known as *pochos* (meaning spoiled or rotten fruit).

The immediate problem of hunger was compounded by debt. Even before the food subsidy program ended, the termination of other government subsidies had made small rural farmers increasingly indebted to the national bank. During the Green Revolution, the Mexican government gave bags of chemical fertilizers to small farmers, who rejoiced because their yields immediately multiplied. Farmers soon abandoned traditional agricultural techniques like the planting of the "three sisters" (corn, beans and squash) that enriched the soil through their rotations. With the increased yields, farmers disdained the methods of corn cultivation that Mexicans themselves originated when they began domesticating corn's ancient ancestor, *teocinte*, in the Central Balsas River Valley 9,000 years ago.

Embracing the technological fix offered by the Green Revolution, modern Mexican farmers quickly became dependent on the habit of "inputs" made from petrochemicals. Increasingly, small farmers relied on expensive fertilizers and pesticides to grow their crops and kill insects. Unfortunately, what they produced appeared cost-effective and marketable only because their manufacturer, the government-owned Pemex, subsidized these chemicals so that they could be given away during the early trial period. Joel Simon was the first American journalist to write at length about Mexico's environmental devastation. He observed that once subsidies were replaced by low-interest government loans, small farmers quickly got into serious debt:

> Three years after they first began to use fertilizers, corn yields...dropped to their original levels. Not only that, if farmers did not add fertilizers, the corn would not grow at all. By the early 1980s, with their debt continuing to grow... people...were caught up in a cycle of dependence. More and more were going to [work in] the United States to...pay off their debts.... The situation had deteriorated to the point where the cost of fertilizers required to produce a marginal corn crop often exceeded the value of the corn itself.[13]

In addition, a related problem had been developing for many years. Long before 1982, rural farmers supplemented their income by exploiting unsuitable, government-owned lands—forests, hills, arroyos, unsettled countryside—until these gave out due to erosion, salinization or desertification. This became a serious ongoing problem. By the time the UN turned its attention to this problem in the late 1990s, about one third (50 million acres, or 20 million hectares) of all Mexican farmland had been severely eroded, while much more (86%) had eroded to some degree. In Tlaxcala, the smallest Mexican state, whose economy relies on rain-fed corn production, half of the state's arable land was destroyed by erosion amidst predictions it would soon become real desert. In the Mixteca region (of Oaxaca state), 70% of all once-arable land is now also ruined. Mexican farmers have a saying that describes the exhaustion of their soil. "The land," they say, "no longer gives."[14]

And, as the land began to give out in the 1970s, more and more peasants learned to survive by working as farm laborers for larger growers who had water, who could afford fertilizers, and who produced crops for the international market. Often too, unemployed rural farmers migrated to large cities, where they worked seasonally in laboring jobs like construction. But in 1982, after economic disaster leveled Mexico's quality of life and eliminated the availability of city jobs, drought struck the most productive zones of Mexican agriculture, resulting in a 40% loss of the corn harvest, and adding to the long list of "push" factors that now began driving people—tens of millions of people—out of Mexico.[15]

There have been numerous studies of the "great Mexican emigration," and many of these find its cause in economic factors like

> negative labor demand shocks, labor supply shocks resulting from demographic changes, US immigration policy changes, migrant networks, and importation of cheap corn and other agricultural products following the North American Free Trade Agreement (NAFTA).[16]

Mexican farmers themselves said they were "fleeing to the United States because they no longer could maintain their previous way of life because of climate-driven crop failures."[17] When we search for the reasons that put rural Mexico to flight, we should listen first to the explanations of those who chose the hard and lonely road to the North.

The drought that first began in 1982 was the vanguard of climate change, and it transformed changes in rainfall, presenting a serious and continuous challenge to local farmers throughout Mexico. By 2012, this drought had expanded out of the tropics to encompass 80% of the United States and destroy what had promised to be the biggest American corn crop ever.

In the Mexico of 2013, heat and drought are so severe that the populations of indigenous hardy species like the distinctive tiny brown mesquite lizard are collapsing.[18] Ongoing drought exacerbates the problem of radical declines in the levels of Mexico's aquifers. As bad as the current water crisis is, it will only get worse. Rainfall changes are having disastrous impacts in already dry states like Tlaxcala and Zacatecas. A former corn farmer in Zacatecas describes the conditions that forced many like him to leave the land and find work elsewhere:

> Ordinary peasant farmers found themselves out of work and short of food. In a good year I can grow four tons of corn and maybe three and a half of beans. But with the sparse rainfall we had this year, I harvested less than a ton of each. Six months of work, plowing, sowing, weeding, fumigating, and that was all I brought in.[19]

Migration to the United States

In the 1980s, massive migration to the United States became the only reasonable course available for rural Mexicans who—due to environmental devastation—would have starved if they stayed in place or who—due to economic devastation—would have done little better in a large Mexican city. In Mexico, these environmental forces are accepted and widely discussed, but in the United States, the same issues are invariably glossed over when American pols play to voters' fears by sounding alarms about how an overwhelming wave of migration is changing the ethnic composition of the United States.

It is true that Latinos are now America's largest ethnic minority and that sometime in this century they will become a majority, outnumbering all other American ethnicities, including Caucasians. This is partly because Mexico is not the only source of Latin environmental migration. The horsemen of financial collapse and environmental devastation have long since visited Central America. Political upheaval and economic chaos impoverish the citizens of Guatemala, El Salvador, Honduras, and Nicaragua, *and* these countries share a geographical rain shadow that is steadily worsening as a result of climate change. Over 8 million residents of Central America's "drought corridor" suffer from chronic malnourishment because rains no longer appear during the planting months of May to August, so crops fail annually.

Since the 1980s, residents of Guatemala have fled to southern Mexico in a wave of migration that is now well-documented in films like *Sin Nombre* (2009). Despite Mexican efforts to repatriate them, between 50,000 and 100,000 unregistered Guatemalans were living in Mexico in the early 1990s, and that was *before* conditions in Central America worsened.[20] By 2002, the rate of chronic malnourishment in Guatemala was 48%, the highest in Central America; the UN sent emergency relief to feed 6,000 Guatemalan children in mortal danger from starvation.

In addition to the Guatemalans fleeing famine in their county, many more Salvadorans also pass through southern Mexico.[21] A famous song by the popular Norteño band, "Los Tigres del Norte," recounts the journey of an undocumented Salvadoran refugee fleeing 5,000 kilometers through Guatemala and Mexico into the United States. "Three Times a Wetback" ("Tres Veces Mojado") has become an anthem for modern Latino migrants, no matter their circumstances or country of origin:

> Para muchos no hay otra solucion
> Que abandonar su patria tal vez para siempre.
> [For many there is no other solution
> than to leave their homeland forever.]

Reaction in America

By the late 1980s, some white Americans in the Southwest had already responded to such massive migration from Mexico and Latin America with outrage and alarm. In 1983, *Time Magazine* warned that Los Angeles was being "invaded [by a] staggering influx of foreign settlers." [22] I remember reading this piece in the shady atrium of the juice bar at the old El Centro market. There was no mention of L.A.'s Canadian community, which at 300,000 was double the size of my small hometown. The "snowflakes" or "snowbirds," as SoCal's Canadian community is sometimes called, were dwarfed by the influx of millions of newly arrived Latinos. While I enjoyed Mexican food and the festival of accents that fills the city's public spaces, the vitriolic rhetoric that targeted these migrants made me increasingly uneasy as I drove through the hazy, palm-lined streets of Silverlake in an old car, listening to salsa stations and pretending I was a tourist visiting Latin America.

But others were not charmed by the Mexican invasion. After the Chiapas Revolution sent a large surge of immigrants across the northern border, in 1994 Californians overwhelmingly approved Proposition 187, denying costly state education and health care benefits to the swelling numbers of illegal aliens. At the same time, millions of illegal immigrants finally became eligible for citizenship under the federal amnesty program passed by Congress in 1986. Across the country, California's Proposition 187 scared Latinos witless; fearing that the conservative backlash in California might become a national scare, eligible American Latinos rushed to become US citizens.

From 1982 onward, Latino immigration represents the largest influx of any single wave of immigrants to the United States in the course of American history. The shift from a trickle to a wave of migrants was sudden and ubiquitous. It challenged America in a variety of perplexing ways. Still, one image stands out among the rest. In 1983, as the influx began, police discovered Rita Quintero, a woman in her early thirties, digging for food in the trashcans of Johnson City, Kansas. Rita, now about 60, is a Tarahumara Indian from the mountain country near Chihuahua. She speaks no English or Spanish. Her native language, Rarámuri, belongs to the same Uto-Aztecan family as other native languages of Kansas (Ute and western Shoshone), but it lacks an alphabet and is not widely known outside of Mexico.

Rarámuri (the Tarahumara name for their language) is substantially different from Uto-Aztecan tongues of the Southwest. In the 1980s, most of the 70,000 or so native speakers lived in rural Chihuahua (where there is now a Tarahumara Internet radio station dedicated to their language and culture). Unfortunately, by the early 1990s, climate change forced the Tarahumara men to move to Ciudad Juárez to seek a different way of life. Traditionally, they were pastoral nomads who practiced *transhumance*—seasonal migration between Chihuahua's lowlands and uplands. But with sharp declines in annual rainfall, the Rarámuri's seasonal pastures failed, so new income sources had to be found. Logging was a growth industry after NAFTA took effect (Mexico lost 7% of its trees between 1994 and 2005). The Rarámuri simply stopped herding and became loggers, felling the trees in the forests of the Sierra Tarahumara mountains that surround Chihuahua.[23]

Before any of this happened, Rita Quintero engaged in a personal migration that took her quite a bit further afield. She is, after all, an Omugi, or female "Tarahumara," a people renowned for their athletic prowess, including the ability to run 200 miles through canyons over a period of several days as part of the Aztecan kickball game *Rarajipari*. (The tribal name, Rarámuri, actually means "those who run fast.") As the crow flies, Rita crossed a distance of

800 miles, covering four states—barefoot—before arriving in a low-rise prairie town of about 1,200 residents, well away from any of the main roads of southwestern Kansas.[24]

Although there is now a small population of both Native Americans and Spanish-speaking Americans who live in Johnson City, at the time no one recognized Rita's language. Portuguese was the police department's best guess. Under interrogation, Rita appeared to be indicating that she had come from the sky, when what she actually meant was that she came from the mountains around Chihuahua. The Johnson City police decided she was mentally disturbed; she had the benefit of a court-appointed attorney and was then committed to Larned State Hospital—where she languished for the next 11 years. During this time, she would sing tribal songs and perform folk dances to occupy and amuse herself in isolation. It was probably for this reason that she was diagnosed as schizophrenic. Then, by chance in 1994, Miguel Giner, a social worker in Great Bend who speaks Rarámuri, heard about Rita. She was released in September 1995 and returned to Chihuahua, where she now lives an old white-haired woman who wryly confesses to hating Kansas.[25] Rita's life became the basis for a famous Mexican play that is still all but unknown in the United States. "La Mujer que Cayo del Cielo" ("The Woman Who Fell from the Sky") was written by one of Mexico's best-known playwrights, Victor Hugo Rascón Banda. It is still unavailable in English translation.

The reality of the migration Banda has documented since his first play, "Los Ilegales" (1980) is far too big for many Americans to accept comfortably. As the wave broke in the early 1980s, it was a very emotional topic—even in academic circles—so I avoided talking about it to my American friends the way one avoids talking politics or religion. I was an illegal alien at the time and vulnerable to the whims of my hosts, although, it is very true that a white English-speaking male hiding in a private university setting was much less visible and much less vulnerable than a brown, Spanish-speaking transient from Central America trying to find work. Americans, I found, are generally very well disposed toward Canadians, whom they still seem to regard as country cousins who just couldn't read their invitation to the family reunion back there in 1776. Whenever I travel south of the 49th parallel, I find little ill will toward a former fellow-colony. Nonetheless, when visiting America in the 1980s, I cautiously held my peace about the "immigration crisis." It was obvious to me that something very important was happening and that it would affect the structure, politics and culture of my adopted home. At the time, I simply opened a mental file on the subject and began to look for reliable sources of information.

Many of California's 37 million people are Latino, of course, but 8% of the state's current population is known to be illegal immigrants (most of them from Mexico). This number increases when the economy is good, and more or less holds steady during economic hard times. Meanwhile, hundreds of thousands of white and black middle-class Californians are leaving the state every year. This deficit outmigration—more people leaving than arriving—began as California itself experienced an economic downturn in the early 1990s.

Since the 1980s, quite a bit has been published in English about the Mexican migration, so we do have some immigration figures, although, since they document illegal immigration, they are often only loose estimates. The current US population hovers at around 314 million people, while the population of Mexico has reached about 112 million. In the year 2000, the US census reported there were 35 million Latinos in the United States (up from 14.6 million in 1980, and up again from 22 million in 1990). Still, official figures describing the size of this emerging minority are inevitably misleading. The known Latino population in the United States includes at least 18–20 million people of Mexican origin who are here *legally*. But there are estimated to be at least 12 million illegal immigrants, and most of these are from Mexico. Undoubtedly, this last figure includes the 6 million Mexican immigrants who applied for foreign identity cards between 2000 and 2006.[26]

It's a good guess that people who applied for these documents are illegal aliens because this form of identification was designed by Mexico's brilliant foreign minister, Jorge G. Casteñada, to help Mexican workers remain in the United States, and there is little other reason for Mexican citizens in America to need these identity cards. These 6 million are only a fraction of the enormous and continuously fluctuating number of illegal Mexicans living in the United States at any given moment (this number includes the "*pochos*"—children born to Mexicans parents living in the US). The Pew Research Institute guesstimates that the total number of Latinos in the United States is 48 million, or 16% of the current US population, and the vast majority of these are Mexicans. To put this in perspective, the current population of California, America's most populous state, is 37 million. Canada, America's closest economic partner and the second largest (in area) country in the world, has 34 million people. So, in all likelihood, the number of Mexicans (and certainly Latinos) who now live as foreigners-in-America is greater than total number of people living in California or all of Canada.

The current population of Mexico is nearly 112 million. The fact that, as a matter of their daily routine, nearly one third of the entire population of Mexico eats, sleeps and works in the United States disturbs many Americans—and has for some time. Because they are immediate neighbors, Mexicans used to return home regularly. For this reason, they did not assimilate readily to American culture any more than do the half-million or so seasonal "snowbirds" who leave Canada to enjoy the warmth of Arizona, California, Florida, Nevada, or New Mexico for several months every winter.

Unlike the snowbirds, who are quite difficult to spot, Mexicans are visibly resented throughout America. The recent wave of narco-terrorist attacks in border-states like Arizona has definitely increased this resentment. But despite a lukewarm welcome in America, the drug-related social and political implosion of places like Ciudad Juárez reinforces the determination of ordinary Mexicans to flee to the United States. Called "Murder City" by its own residents, CJ has become the symbol of how bad things will soon become throughout Mexico. William Booth, the *Washington Post*'s Mexican correspondent, describes the city eloquently. He writes, CJ is falling apart. Years of neglect have left streets mined with potholes. The parks are ruins, the playing fields nothing but weeds, the once lively cantinas shuttered. There are few schools on the poor side of town, where barrios of cement-block houses have been abandoned by fleeing residents, who went home to their villages or crossed illegally into the United States. The city's business elite have moved to El Paso. Although the poor have always struggled here, they are now hungry.[27]

Changes in the Mexican View of America

When Americans bother to reflect on the causes of Mexican migration, they usually attribute the movement to Mexico's economic collapse which, once again, began decisively in 1982. But it's much more accurate to think of the swelling wave of Mexican migration as the result of a one-two combination of punches in which *first* the economy, *and then* the environment connected with Mexico's undefended chin. As this happened, many Mexicans began to revise their image of the United States. According to sociologist Alejandro Portes, a Cuban exile who has devoted his life to studying Latin migration, after 1982 America came to symbolize "the land where the problems of Mexico [could] be solved," where "the benefits of an advanced economy promised but not delivered by the Mexican development plan [could] be turned into reality."[28]

Over the next 30 years, America gradually became the symbol *and* the site of Mexican economic salvation. Documents in English about the experience of Mexican migrants themselves are hard to locate, but the symbol of migration has become an intrinsic part of the Mexican cultural mindset. One rare English language work exploring migrants' motives and experiences is the eloquent and recent volume *Crossing with the Virgin*, which describes the genuine universality of Mexican migration:

Not only very poor people were crossing the desert. We would meet successful people who had fallen on bad financial times. We would meet people like my godparents who had strong businesses: naturopaths, shop owners, corporate executives and bankers. We would meet funny, smart, hardworking ranchers and farmers.... Not everyone in Mexico is queuing up to come to the United States. But many are. Most people who come are in crisis.[29]

Gradually, as Mexican attitudes to *el Norte* changed, it became the site of economic salvation—and anti-US sentiment progressively disappeared. Apparently, a high percentage of Mexicans now have "a favorable opinion of the institutions of American society."[30] It is remarkable that in response to national polls conducted in 1991 and 2002, nearly 60% of Mexicans said they favored political union with the United States if that would improve Mexican living standards.[31]

Cynics will say that Mexicans have only acquired their more appreciative attitude toward the United States because of Mexico's terminal economic and environmental chaos. No doubt they're partly right, but this may still be a very good thing. The radical change in recent Mexican attitudes toward the United States is a marked improvement over the deep suspicion, mistrust and hostility that accompanied Mexican nationalism since the humiliating Mexican-American war of 1847. Still, what fascinates me most is the knowledge that this combination of economic and environmental collapse is always at the root of social orders whose spirals of decline bring on a Dark Age or "critical crisis period." According to the founding editor of the magazine *Nature and Culture*, it is at these moments that "environmental conditions play a significant role in determining how societies...are reorganized."[32]

The Pattern of Social Collapse

Many people are familiar with Jared Diamond's notion that with increasing complexity human societies tend more and more toward systemic collapse. In his trilogy on world ecological degradation, economic historian Sing C. Chew quantifies Diamond's observations by tracing connections between the environment and the collapse of expansive economies immediately prior to the onset of each Dark Age. In each of Chew's case studies, economic collapse is accompanied by the simultaneous collapse of the natural environment that had sustained the society. Chew claims—and it seems sensible—that all economies are dependent on the environments that sustain them, so economic collapse is characterized by a loss of the environmental carrying capacity for human expansion and progress. Political scientist Stephen Mumme feels this combination of conditions exactly describes Mexico's dilemma after 1982:

> If, in the 1980s, crisis was the most common adjective used to describe the Mexican economy, much the same could be said of its environmental predicament. Forty years of rapid industrialisation…had taken a devastating toll on its natural resources and the health of its citizens.… It is hardly an exaggeration to say that the deterioration of Mexico's environment has been comprehensive and on a magnitude with few rivals.[33]

Mexico's Environmental Collapse

The destruction of Mexico's arable land has worsened considerably in the decades following "la crisis." A study concerning the sharp decline of available water throughout Mexico by Scott Whiteford and Roberto Melville, explains what the disappearance of good soil means for food production in Mexico:

> By 1988, more than 400,000 hectares had gone out of [farm] production because of salinization. This is equal to 1 million tons of food grains…enough to satisfy the basic needs of 5 million people.… Another 100,000 will be fallow by 2005 because of salinization and water-logging. This increasing agricultural contamination affects the surface water that feeds into lakes and aquifers.[34]

Mexico's ongoing war on its own topsoil has virtually ended the indigenous production of corn, beans and squash for small-holdings
farmers—the majority of all agricultural producers in Mexico.
Utter devastation of the Mexican environment is the result of the
rush toward industrialization in all potentially commercial fields.
With the single exception of China, Mexico leads the world in industrial damage to its environmental foundation. The comparison
to China's infamous pollution and degradation is no exaggeration.
In 1995, the UN released this damning overview of Mexico's environment:

> The ecological crisis [is] seen in the net reduction of [Mexi-
> co's] forests at a rate exceeding a million acres annually…the
> reduction of the Lacandon forest zone by 70 percent since
> 1950, and the disappearance of thousands of species of fauna
> and flora in a nation with one of the world's highest levels of
> biodiversity. The threat from environmental pollution [is]
> evident in the severe degradation of its two most celebrated
> natural lakes, Lake Chapala and Lake Patzcuaro; the con-
> tamination of over 60 percent of its rivers; severe oil spills
> along the Mexican Gulf coast, damaging national fisheries
> and aquatic life; inadequate sanitation and sewerage facili-
> ties in more than half of Mexico's municipalities, both large
> and small; the virtual absence of hazardous waste disposal
> facilities throughout the nation; and, perhaps most noto-
> riously, the venomous air pollution blanketing the world's
> largest urban area, Mexico City, and one of the highest rates
> of pulmonary disease on the globe.[35]

The industrial assault on Mexico's environment is so comprehensive
that, in 1995, the UN sounded an alarm about the impact that imminent environmental collapse would have on the country's population. They warned that the lowest-income groups would soon
experience a radical decline in living standards; that the countryside
would continue to become impoverished; that land would become

less productive; and that corn and bean yields, which are the staples of the Mexican diet, would continue to decline until there is "widespread hunger in the birthplace of the green revolution."[36]

In 2005, drug-cartel violence in Mihoaćan signaled the beginning of an unending assault on Mexican social stability that has so far killed about 50,000 people.[37] But, well before this, Mexicans had many strong reasons to leave for the United States and very few reasons—other than family and friends—to return. In the coming years, if economic health gradually seeps back into the United States, many more Mexicans will leave because the country has become chaotic and dangerous in addition to being very short of food, water and work. So far—and it will not remain this way—Mexico has been able to survive the onset of a Central American Dark Age (or what is now usually called "failed state status") by substantially reducing its population and importing large amounts of foreign currency to support those who remain in the country. Sending Mexicans north to work for their wealthy neighbor leaves fewer mouths to feed in Mexico, and Mexicans working in America send money home to sustain their loved ones. No wonder the Mexican government does everything it can to encourage northward migration.

In the United States, the impact of Mexico's one-two economic-environmental punch has been profound. Economically and socially, Mexico may have been knocked down by these punches, but, as it fell, much of Mexico actually landed in America. The 9 billion or so dollars of remittances that Mexican Americans sent home every year before the economy collapsed in 2007 would go a long way to shoring up America's recessional unemployed. Many out-of-work Americans are angry that jobs in the United States go to illegal Mexican immigrants whose "criminality" prevents them from launching labor actions. But the so-called criminality of illegal immigration is a fiction, as Professors Gregory and John Weeks explain in their powerful book *Irresistible Forces: Explaining Latin American Immigration to the United States*:

It is common...for undocumented immigrants to be called "criminals" who have engaged in "unlawful activity" because they have broken the law by entering the country without authorization. But, in fact, it is not a criminal but rather a civil offense—the same level of seriousness as minor traffic violations. [38]

Nonetheless, Mexican illegals are powerless to prevent ruthless employers from undercutting America's minimum wage. Ironically, the illegal cheapness of illegal migrant workers is what makes them so necessary to America's economy. There is no need to outsource jobs when you can pay your workers the same wages they would earn in a poor nation in the developing world. This is why illegals continue to find jobs in America today.

Illegal immigrants serve the economy of two countries with fundamentally different interests. In Mexico, workers' remittances from America are vital to the nation's economic survival, and the Mexican government has recently taken the extraordinary step of filing an *amicus brief* in a lawsuit (Friendly House, et al. v. Michael B. Whiting) to overturn SB 1070, Arizona's controversial anti-immigration law. (An amicus brief is an attempt to lobby a court by a friendly third party—*amicus curiae*—whose interests are not formally represented in a specific case.) The Mexican brief asked that the court declare Arizona's new law unconstitutional, describing the issue of immigration to America as one of

> great importance to the people of Mexico, including the almost 20 million Mexican workers, tourists and students lawfully admitted to the United States throughout 2009, [to] those already present or who will similarly be admitted to the U.S. in the future, and [to] the countless millions affected by international trade, immigration policies and drug violence.[39]

In the summer of 2012, the Supreme Court upheld the legality of SB 1070. (Though, as of the time of this writing, the Obama administration is working to frame a comprehensive immigration law that will permit illegal immigrants to work toward US citizenship.) Nonetheless, Mexicans themselves feel that they must continue migrating to America because there are few other options. Despite the obstacles in their path, they will continue to come in greater and greater numbers to escape the growing shortages of food, water, work and security.

The "Urban Stain" of Mexico City

Human beings have the advantage of mobility, and the option of migration has always been our species first response to climate change. Other species are not so fortunate. Around Mexico City, the *ajolote* (ah-ho-LO-tay) a 9-inch-long salamander once believed to be the incarnation of *Xolotl* has all but disappeared from the polluted waters of its only environment, the canals of Xochimilco. The ajolote is a pivotal image of the Latin soul (and the central topic of the powerful story "Ajolotl," by Argentinian expatriate writer, Julio Cortázar, also the author of *Blow Up*). Mexicans blame *la mancha urbana*, for their salamander's decline. But what they call the "urban stain" is only part of the profound environmental devastation that Mexico has suffered for over half a century.

Shortages of food, water and work—as well as the urban stain— are the major push factors that drive ongoing migration from Mexico into the United States and reinforce the decision of Mexicans who have already reached America *not* to return to Mexico despite the hardships of recession. Lander Mondragon summed up the determination of Mexican immigrants to remain in the United States when he experienced a work shortage in Atlanta in 2008 "I ain't got nothin' to go back to in Mexico," he said. "This is my hometown now."[40] A bitter measurement of Mexican immigrants' distaste for their birth country is the rise in the number of deaths that occur during illegal crossings. By July 2010, the number of people dying of dehydration or hyperthermia in the Arizonan desert exceeded 140, the greatest annual number in this century. As a heat wave parched Mexico and the American Southwest, it killed many of those desperately trying to escape the devastation of their country by moving to a better life.[41] Since the heat-island effect prevents night-time cooling, migrants crossing the desert get little respite from daily highs of 109°F during the three days it takes to cross the border by foot. About 500 people will die this year while illegally migrating to the United States from Mexico. There is no known force that can stop this migration or reduce the number of the dead. Economic destruction and environmental collapse are squeezing Mexicans into the United States as though they were toothpaste leaving a tube.

By mid-century, much of America will be Latino, but the acceptance of Latin culture as an intrinsic part of American culture began when Mexican immigrants to America finally organized to reject the Sensenbrenner Bill in 2006. At last, Mexican-born Americans were taking an interest in political life in the United States and presenting themselves as a voting bloc to be courted by the major parties. As professor Leo Chavez puts it: "The immigrant marches of 2006 were not one event but many."[42] For the first time, Mexican and Latino activists were galvanized into action in the United States. Their ad hoc coalition spread across the country via the Internet, radio and other mass media that catered to immigrants and Latinos. All of these prompted Latino participation in peaceful demonstrations across the country. There were calls for work and school boycotts and for moratoriums on buying and selling in order to demonstrate the size of America's Latino minority and its collective financial muscle. The response was remarkable. In 2006, civic demonstrations against the Sensenbrenner legislation by Mexicans and Latinos crisscrossed the country from March until May. Other ethnic groups soon joined in. Many were American Catholics from large urban centers of Catholicism. Cardinal Mahoney of L.A. and Mayor Richard M. Daley of Chicago spoke forcefully, favorably and often, about the role of immigrants in America and the rights they should be accorded. There was also participation by Korean and Chinese Americans.

These demonstrations were larger than the historic civil rights march on Washington in 1963 or even the very largest of the anti-Vietnam demonstrations in the late 1960s. They included a march and demonstration in Chicago on March 10th, estimated to involve 300,000 people; a march and demonstration on March 25th, involving 500,000 people in downtown Los Angeles and similar marches in Phoenix and Charlotte; marches in Oakland, San Francisco, Fresno, Yakima, Washington DC, Phoenix, Detroit, Columbus, Ohio, Houston, Woodbridge VA, Norwood MA, and Longmont CO; demonstrations on April 9th of about 400,000 people in Dallas, with similar but smaller marches in San Diego, Miami, St. Paul, Birmingham, Des Moines, St. Louis, Salem and Boise; on April 10th, rallies in Washington DC and Los Angeles with about 500,000 marchers each and other large rallies in Phoenix, Houston, Omaha, Boston, Atlanta and many other cities; and, on May Day, there were demonstrations in most large cities across the country, the largest being those of Los Angeles and Chicago with about a half million demonstrators each.

The symbolism at these events is fascinating. A large number of Mexican flags were noticeable and, of course, these indicated the heritage of many of the marchers. But in a remarkable show of solidarity, demonstrators also sported white shirts and tops that gave viewers the impression of an enormous army united in a common and decent cause. Many, many people also carried American flags—clearly intended to send the message that these demonstrators no longer saw themselves as foreign workers. They were claiming their rights and status as taxpayers and as an integral part of the US economy. As one marcher put it "A lot of us…broke the law to get here. That doesn't mean we don't love America."[43] Or as another protestor said:

> I'm legal. But if I try to help someone who has no papers [according to HR 4437] I'm a criminal. For years, I was very quiet. I only worked and paid taxes. Now it's necessary to protest.[44]

In Washington, the political reaction to this overwhelming display of solidarity was abrupt and immediate. Republicans in the House of Representatives started to back away from the most extreme measures of HR 4437, including those that would make it a felony to be in the United States without documentation or that would make it a felony to help an "indocumentado." Some Republicans tried to shift the blame onto their opponents, claiming that the Democrats had somehow been responsible for removing the offensive criminal provisions from the bill.[45] But ten years of activism and integration had left the demonstrators with greater political acuity than the busy residents of South Gate in Los Angeles who had been manipulated for many years by old-style political bosses like Albert Robles.

Among these newly politicized Latinos, there was a realization that their greater participation in the processes of American political life would result in improved conditions and greater respect. So, the spring of 2006 should be understood as a turning point in American electoral politics. It was the moment when America's Mexican minority came of age, becoming an important voting bloc in American national and state level politics. This transition underlies Sonia Sotomayor's ability to win GOP approval for a seat on the Supreme Court in 2009; very few senior Republicans were willing to risk opposing her nomination.

The arrival of a Latin voting bloc means they have become players in national politics; this, of course, has great significance for America's future. Demographic projections have it that by 2050, a majority of all Americans will be Latino, and most of these will be of Mexican ancestry. But in addition to changes in the complexion of America, there is a much more vital lesson to be learned from the great Mexican migration and the integration of Latinos into American culture. It is this: *when economic collapse is accompanied by environmental collapse, the resulting destruction of any region's carrying capacity initiates massive human migrations.*

Such migrations are measured in tens of millions.

Final Word

Few people would describe California's budgetary crisis as the kind of environmental implosion that struck Mexico in 1982, and yet—just as there was in Mexico 30 years ago—there is a pressing environmental dimension to the state's predicament. In the Golden State, the volume of available water is shrinking on all fronts due to the same climate changes that dried out northern Mexico. As the earth's tropics expand toward the poles, previously fertile regions are becoming increasingly arid. Already, California's dense population, advanced industry, and huge agribusiness require more water than is available, and, because its financial crisis has already lasted for decades, this region (that has the eighth largest economy in the world) can no longer afford to buy increasing amounts of water elsewhere. Despite the best efforts of Governor Arnold and Governor Moonbeam, little is being done to resolve this catastrophic lack. Today, like Mexico, California teeters on the brink of becoming a failed state as the result of a devastating financial crisis that will soon be compounded by a disastrous reduction in available water and the loss of its agribusiness.

Outmigration is the obvious long-term solution to the life-threatening scarcity of such a vital resource as water. Already, at 270,000, California has the second highest number of moving vans leaving the state annually.[46] And that number represents only the number of middle-class families leaving; outmigration has depleted the state's wage-dependent urbanites in both the lower and middle classes.[47] Consequent to the 2010 census, California may soon lose one of its 53 congressional seats. Of course, these migrants are not "environmental refugees" and their decision is not a "climate migration." But if climate change occurs during the kind of financial downturn that already causes people to leave a region, its crushing expense will then tip the economic scales into financial ruin. Historically, a combination of financial chaos and ecological devastation has initiated every known example of "failed state status."

Today (in the early autumn of 2012) California stares into an abyss similar in kind and cause to the double-barreled catastrophe that assailed Mexico in 1982. Since 2003, the outmigration from California's urban centers that began among low-skilled workers across all ethnicities in the mid-1990s, now extends to its middle classes. Former Los Angeleño, Candice Reed, a journalist and author is one of the most vocal of these. In late summer of 2009, Reed published a farewell to California as "Dear California, I'm dumping you" in the *Los Angeles Times*:

> Dear California…
> You've totally lost perspective…. I'm sinking into depression!
> We can't pay our bills…the phone is ringing off the hook
> with creditors…. Children…are losing healthcare, more
> than 766,300 Californians lost their jobs…last year…we're
> at the top of the foreclosure charts. You need to change and
> you refuse to admit it…. There's no doubt I still have feelings
> for you but…. I lost my job in the newspaper industry and
> my house is being sold under duress. I want out. I'm leaving
> you…and you might as well know the truth; there's another
> state…. I'm falling…for.[48]

In plainer terms, the quality of life that once made California synonymous with paradise is in decline. Like Candice Reed, middle-class Americans are finding opportunities for employment and more affordable lives with better quality elsewhere. Reed has moved to Chelan, Washington, where she now works for a small local paper while her retired husband has taken a job he loves laboring in a local vineyard. The growing exodus from California in which the Reeds are participating is the result of an economic crisis brought on by the government's attempt to sustain a high societal quality of life without levying additional taxes. An American *state* cannot declare bankruptcy, but in 2012 many large Californian *municipalities* discovered they could, and they did exactly that. Peter Schrag, author of *Paradise Lost*, the book that first warned Californians about the state's nearly inevitable decline, writes:

California...had been coasting on the capital investment of the 1950s and 1960s and had...been disinvesting by letting that infrastructure deteriorate.... In 1960 California spent nearly $1.50 per person for infrastructure. In the 1980s and 1990s, it spent roughly $0.25. California's backlog on infra-structure—the schools and other public buildings, roads, water and sewer systems that needed to be built, repaired or modernized—was conservatively priced by the California Business Roundtable in the late 1990s at $90 billion, by a state commission in 2002 at $100 billion, and by others at considerably more.[49]

In addition to the rapid downturn of California's economy and in-frastructure, there is in store a state-wide environmental crisis of proportions *unequalled before* in any of the United States. Finding enough water for 37 million Californians is, of course, already a ma-jor challenge. There's not much to be found elsewhere and, anyway, California can no longer afford to buy it. In addition to the problem of a general lack of potable water that will already squeeze millions of people out of the state, there is also the problem water quality. In recent decades, salinity has become a serious threat. Most Cali-fornian water comes from the aquifers or from the lower Colorado River. In the past decade, because of aquifer depletion and evapora-tion at reservoirs, the concentration of salts has increased in both sources (because the same amount of minerals is now dissolved in much less water). California's famers are forced to irrigate fewer acres with the same volume of water, to switch to more salt-tolerant crops, to install expensive tile drains, or to somehow obtain more water to produce a dwindling volume of crops. In 2001, the problem became especially obvious for farmers in the Imperial Valley, the centerpiece of Californian agribusiness. In *The Great Thirst*, a book about the history and future of water crises in California, Norris Hundley describes what's in store:

> The Imperial Valley has been especially hard hit, pouring millions of dollars into a struggle to control salts that, barring

some unexpected technological breakthrough or infusion of new water, will inevitably be lost. If that happens, the valley will be abandoned, thus following a pattern established by many earlier civilizations stretching as far back as Sumer in the third millennium BC.[50]

In the coming years, droughts, heat waves, and increasingly large forest megafires (like the ones near Bishop, Lake Nacimiento, San Bernardino, Ventura and Santa Cruz) will intensify the state's already irreparable economic devastation while reducing its carrying capacity and making California—especially Southern California—a truly miserable place to live. The dream is genuinely over. This is the beginning of the end. As I write this, I am very sad because I love California but have only been able to return to it twice in the past 20 years. In the meantime, it has been ruined by greed and mismanagement.

Without any help in sight, California is now unable to cope with any major crisis—a megafire, an earthquake, a drought—so ongoing climate change can only continue to kick the state, and keep it down in the coming years. I remember my first day on the beach in Santa Monica in late August 1980. It was so beautiful I thought that I would stay in SoCal forever. But later, we had kids and no health insurance so we left for a more affordable life in my native Canada. Now, the beach paradise I once loved is gone.

California today is like that song-of-a-girl you were going to find the nerve to dance with before the party ended. Ever since, you've felt the harsh disappointment of not having asked her. And then, when you run into her at a class reunion, life will have weighed as heavily upon her as it has upon you. The beauty that you thought a solid reality is yanked out of your memory. With the girl, it's simply a matter of life and aging. With California, it was the result of mismanagement, stupidity and greed.

Time to move on.

CHAPTER 3

Runnin' Dry

Canada, the most water-rich country...in the world, is going to get richer in the 21st century, perhaps as much as 40% richer (wetter) by mid-century. [But] the United States' most booming regions, already on water overdrafts, may lose nearly a third of the water they enjoy today.

Chris Wood, Dry Spring (2008)

We need water with only slightly less urgency than we need air. An infectiously happy scientist on CNN brought this point home in the summer of 2008 after an unmanned spacecraft (the *Phoenix*) landed successfully on Mars. Its robotic arm was programmed to deploy an ordinary router-bit that Jet Propulsion Laboratory scientists found during a lucky trip to the Pasadena Home Depot. With this, the *Phoenix* ground a small hole into the polar ice cap to determine if Martian water had ever melted and then mixed with the surrounding soil. "We found it!" screamed NASA's exobiologist into CNN's well-placed camera. Then he explained this eureka moment: "Water" he said "is the first requirement for extraterrestrial life."

On our world too, water is vital to life, and here conditions are much more favorable than those on the Red Planet. But even here, the freshwater that sustains surface life is a rare commodity. According to NASA, earth's oceans contain 96.5% of the 326 quintillion (18 zeroes) gallons of terrestrial H_2O. Of course, ocean water is far too salty to drink (and desalination is still a costly process). Fortunately, the earth's hydrologic cycle evaporates 473 trillion tons of ocean water and another 73 trillion tons of landmass water every year. Nearly one fifth of this falls back on land as rain or snow. If this naturally distilled freshwater were spread evenly across every continent, the world's precipitation would amount to about 30 inches everywhere, per year. There'd be plenty of clean water to go around, and everyone would be happy. Unfortunately, in this imperfect world, about 50% of precipitation re-evaporates almost immediately, and what's left is never spread evenly across the earth, so regional shortages are inevitable.[1]

To overcome shortages, men have dammed, diverted, and drawn-down upon rivers and lakes on the earth's surface for centuries. These large bodies of open water are not the largest reservoirs of freshwater, but they are the most accessible. We conquered these low-hanging fruits early in our history when our technology was surprisingly sophisticated, though much less powerful. The Upper Mesopotamians and the Sassanid Persians developed a sophisticated method of accessing natural aquifers. They called their invention *qanats*, or chain-wells. And these same people developed the first irrigation systems based on aqueducts that could redistribute freshwater on a national scale. Other civilizations as far-flung as the Romans, the Incas and the Khmer (Angkorians) repeated Persia's formidable accomplishment. From an engineer's point of view, the ascent of man would never have happened if we hadn't learned how to manipulate water resources and adapt the earth's carrying capacity to the needs of a thirsty and agriculturally inclined species.[2]

Water was especially important during the settlement of the American West. During the 19th century, William E. Smythe, father of America's *Irrigation Movement*, theorized that "aridity" was, in fact, a divine blessing because water for irrigation "lay beyond reach of the individual." According to Smythe, the common cause of defeating aridity in America's driest territories created social capital—civilization—by making cooperation a necessity. The enforced cooperation over water, Smythe felt, would make a great civilization bloom in the American West. Months before the beginning of the 20th century, he wrote that the

> organization of men [was] the price of life and prosperity in the arid West. The alternative was starvation. The plant which grew from this new seed was associative enterprise, and we shall presently see what flower it bore.[3]

A hundred years later, there is recurrent drought in 13 American states. In America, as elsewhere, rapid population growth resembles popcorn expanding in a small, hot pan. Today, intensifying demands on our freshwater resources are more likely to create wars than to reinforce social capital. Already our planet is straining with the effort of supporting 7 billion people. By 2050, there will be nearly 3 billion more. Census predictions have it that the United States itself will swell to about 400 million citizens by mid-century. Climatologist Kevin Trenbreth has demonstrated that after a century of little change, drought events rose sharply after 1970, and that, while extremely arid regions represented only 15% of the earth's surface, they have now expanded to 30%.[4] So while the world's population expands, topsoil abuse and climate change are shrinking the earth's habitable lands, including those in the United States. Already, one billion people do not have access to sufficient potable water. The UN now predicts that by 2025, population increases, aquifer depletion and drought will leave two thirds of the world in similarly desperate straits.[5]

Disappearing Aquifers

Because of aquifer depletion and drought, outmigration on America's High Plains is now almost complete. Misleadingly, the area is called a "semi-arid region." That term doesn't mean (as you might think) that the land receives some-but-not-much rain. Instead, it means that levels of rainfall vary sharply across vaguely defined periods of years. If Western farmers didn't already have enough challenges, semi-aridity describes unpredictable climatic conditions alternating between drought and the occasional abundance of water. Recently, scientists have discovered that this tug-of-war between drought and plenty in the West reflects cycles of a powerful interaction between ocean currents and winds that originates in the southern Pacific Ocean.

La Niña

The semi-aridity that results from the El Niño-La Niña phenomenon has kept the population of the Plains low in comparison to that of the water-rich East Coast. But it's the poleward expansion of the tropics under climate change that's the real kicker in the discussion of North America's future droughts. The Southwest, the Midwest and the Southeast have all experienced increasingly dry conditions since the turn of the millennium. Unfortunately, this region's drought is neither temporary nor regional. It will persist through the rest of this century, and its reach is global.

The meteorological phenomenon called El Niño guarantees increased precipitation, but unfortunately its complementary phenomenon, La Niña, has produced a swath of drought across the continental United States from California to the southeast Coastal Plain; drought-troubled states are not just confined to the Southwest. After years of drought, Lake Lanier, Atlanta's main reservoir, contains a supply of water only sufficient for a few more months. The prevalence of La Niña conditions appears to be growing in our century. In fact, the expanding belt of drought that derives from La Niña now stretches around the world, and this first brought it to scientists' attention in January 2003.

La Niña is a meteorological phenomenon of the eastern Pacific Ocean, a mirror image of the El Niño Southern Oscillation (ENSO) effect in the western Pacific. Officially, these terms describe sea surface temperature anomalies of 0.5°C or greater across the central tropical Pacific Ocean. (If the anomaly lasts up to five months, it's called an El Niño or La Niña *condition*. If it lasts longer, it's an El Niño or La Niña *episode*.) El Niño and La Niña are sometimes described as "quasi-periodic" conditions because, while their underlying mechanisms are poorly understood, there is general agreement that they alternate repeatedly, even though they do this in unexplained and irregular periods of two to seven years. In the 20th century, there were 25 El Niño events and only 18 La Niñas.

Under El Niño, sea surface temperatures in the western Pacific *rise*, causing large volumes of water to evaporate. The effect of El Niño on America is often intense storm activity and torrential rainfall from the Pacific Coast throughout the Southwest. During La Niña, however, colder sea surface temperatures in the central and eastern equatorial Pacific evaporate much less water. The most recent La Niña brought with it a far-reaching drought that stretched

> over an extensive swath of the Northern Hemisphere mid-latitudes spanning the United States, the Mediterranean, southern Europe, and Southwest and Central Asia.... As little as 50% of the...annual average precipitation fell in these areas during the 4-year period [1998–2002].[6]

The discovery of the impact of La Niña's oceanic cooling on drought events has now been applied to historical investigations of droughts. This is only possible because the world's navies began recording sea surface temperatures regularly in 1856.[7] Using the navy readings, scientist Richard Seager has demonstrated that the six major Western droughts in recorded history—including the Dust Bowl—were caused by La Niña.[8] The significance of discovering a single, underlying cause for all these droughts is substantial since in the coming century, global warming will probably increase conditions favorable to La Niña. In the 21st century, the Pacific's 20th-century predisposition toward El Niño will be reversed. Increasingly La Niña will be the dominant meteorological anomaly. With it will come massive drought events across the subtropics of the Northern Hemisphere. During this period of intense climate change, the Southwest will stop being a semi-arid region and become a completely arid one.[9]

Outmigration

In the 1980s, Deborah and Frank Popper, a husband and wife research team, observed that the West's population was actually in decline and would probably continue declining until much of the region once again became "frontier" (which is actually a technical demographic term describing areas with fewer than six people per square mile). Frank Popper, an urban planner, began this investigation because he was drawn to the phenomenon of ghost towns. Deborah, a social geographer, had deeper interests. She was fascinated with discovering the "secrets of survival and stability" that made some towns last when others failed. With her research, she tries to explain why

> some small Plains towns [are] dying while others are clearly possessed, against long odds, of the will to live.... What factors make the difference?[10]

The Poppers published their predictions in an infamous essay called "The Great Plains: From Dust to Dust." Basically, their work is a well-formed and quite reasonable argument that Western areas abandoned after a precipitous population decline should be returned to the conditions of the original prairie by reintroducing indigenous species, including American bison.

These ideas had a very innocent beginning. Husband and wife were both very familiar with the demographic information about outmigration from the American West. Stuck in traffic one day on the Jersey turnpike, Frank and Deborah had a spirited argument about what should be done with the vast empty tracts of land in the rapidly depopulating High Plains. By the time they had reached the Verrazano Bridge, a brilliant solution had come to both of them. Generously (and naïvely), they decided to share their inspiration with their fellow Americans.[11]

The High Plains was never a very populous region, but it is nonetheless a remarkable one. While only 2 million (less than 1%) of Americans now live in the states that share the water of the immense Ogallala Aquifer, the region is responsible for 35% of America's agriculture.[12] What happens there affects the food supply of the nation and the world. Colorado, Kansas, Nebraska, New Mexico, Oklahoma, and, to a lesser degree South Dakota and Wyoming, are America's breadbasket. This is only possible because of the "fossil water" under the soil of this semi-arid region. For this reason, the area is sometimes called the "land of the underground rain."

At this moment, most existing freshwater is cycling through one of two kinds of natural reservoirs. Both have been collecting water for tens of thousands of years. The first of these is well known. It's the frozen, crystallized water stored in the world's glaciers, snow-packs, and polar caps, sometimes called the *cryosphere*. The second freshwater reserve is less well known, probably because it's much less visible. "Underground rain," or groundwater, is stored in aqui-fers—vast areas beneath the earth's surface. About one half of all the freshwater in the world is stored (at least temporarily) in frozen form. Another quarter or so is embedded as groundwater in sand or gravel under our feet, where it is neatly insulated from evaporation.

Aquifers come in two kinds. Some, like the one near the Madi-son River and the Old Faithful geyser are accessible to rain or surface water and can be recharged annually. During periods of recurring drought, ongoing aquifer depletion occurs because the aquifer is not recharged. It is because its surrounding aquifer has been depleted by drought that Old Faithful's eruptions are now spaced 90 minutes apart, a full half-hour longer than they were a century ago, when the waterspout was first discovered and named.

The other kind of aquifer cannot be recharged. These are the "fossil aquifers" like the Ogallala on the High Plains (or, rather, un-derneath them). This stock of water is geologically sealed beneath the earth's surface and can never be replenished. Combined, these two types of aquifers contain 30 times more of earth's water than all of its rivers and lakes, which are, in fact, the smallest freshwater collectors (and the ones most susceptible to evaporative losses).

Because the Ogallala Aquifer can never be recharged, water withdrawn from under the High Plains permanently depletes the overall volume of freshwater available to the region. Aquifer depletion is one of the main reasons that the West is going dry and losing its agriculture-based population.

Overuse of Technology

The Ogallala was originally formed 10,000 years ago by the melting glacial ice of the western Rockies. The presence (if not the vast extent) of this groundwater has been widely known since the 19th century. At first, the problem was how to access it. Knowledge of the irrigation system used in northern Mexico would have helped the settlers enormously: *Qanats*, which manipulated a sophisticated appreciation of the behavior of water tables, had been irrigating farms in Mexico for nearly 2,000 years; they were series of chain-wells "cut into the gradual gradient of a hillside."[13] Many of the farmers on the High Plains had European roots, and, because water is plentiful in continental Europe, farmers there had less need for practical knowledge about aquifers. In Mexico, however, it was quite a different story. Chain-wells are part of the Iberian-Islamic cultural legacy that spread from their place of origin in Persia (where they are called *karez*) to the mysterious Garamantian civilization of central Libya. The Moors later established working chain-well systems in Morocco at Erfoud, and then in Muslim Spain at Córdoba, Crevillente and, famously, in Madrid before the Islamic expulsion. From Christian Spain, after the Conquest, knowledge of these wells spread with Catholic missionaries into northern Mexico. Meanwhile, in Anglophone, Protestant America, the technology of chain-wells remained unknown.

Many Americans in the old West subscribed to the "folk" notion that groundwater flowed under the earth's surface in subterranean rivers, like the famous *Lost River* of Appalachia. In fact, this kind of geological feature is quite rare. Still, 30% of the world's freshwater is groundwater, and the vast volume of water known to reside underground created widespread belief in open underground channels or rivers. This was a popular belief of the period and is perpetuated by many myths and legends. In "Kubla Khan," (1816) Samuel Taylor Coleridge described "Alph, the sacred river" running through measureless caverns, but this image itself is as old as the river Styx that winds through Hades. The belief was so widespread, it conditioned the settlers' attempts to forcefully "raise" the water out of its sources.

For me, the difference in these cultural images of underground water resolve into the differences between information derived from science and that derived from myth and ideology. This opposition provides a telling illustration of how our received ideas about nature shape our interactions with it. In the American West, the problem of accessing a seemingly limitless supply of aquifer water flowing through underground streams seemed like a simple pumping problem. In semi-arid countries that use *qanat* technology, in which gravity moves groundwater down a gradual slope into channels that lead to farmers' irrigated fields, the passive nature of the system limits the degree to which farmers can draw upon an aquifer. Unfortunately, the High Plains didn't benefit from *qanats*. Instead, generations of increasingly powerful pumps were designed to draw more and more water out of subterranean sources in much the same way we might put a hose in a river, attach it to a pump and then water a field with the output.

As the conceptual focus on the problem of irrigation shifted toward the technology of pumping, people increasingly forgot the advice John Wesley Powell gave to Congress in 1879. In describing the American West, the one-armed Civil War veteran who had led the first expedition down the Colorado, often used a theological metaphor that he inherited from the Mormon settlers of Utah. Powell admired Mormons because they had skillfully adapted to the conditions of a harsh land. While Powell noted that many parts of the "Arid Region" could be "redeemed" by irrigation, he advised Congress against overpopulating and overtaxing local water supplies. Powell told them that "only a small portion of the country can be redeemed. The irrigable tracts are lowlands lying along the streams."[14]

Advocates of the *Irrigation Movement* disagreed completely. The urge to populate and civilize the frontier had caught the nation's

imagination. Powell, a war hero and explorer of national reputation, fought them, campaigning for moderate growth in the West.
In 1893, he visited Los Angeles and warned the International Irrigation Congress against plans to irrigate America's West: "I tell you
gentlemen," he told the assembly, "there is not sufficient water to
supply the land."[15]

The "Irrigationists," however, prevailed. It's a cruel irony that
Lake Powell, the second largest reservoir in the United States, bears
the name of the explorer who devoted so much personal effort to
preserving America's Western wilderness and defeating the cause
of irrigation. John Powell's stand against Western expansion was
swept away in the nation's fervor to fulfill its continental destiny.
By the 1890s, the Indian Wars had ended, and land in the West
was cheap and empty. Publications like Joseph Bristow's *Irrigation
Farmer* (1894) or William Smythe's *Irrigation Age* (1896) promoted
widespread irrigation and settlement of the entire region. Innovative new pumps were developed and then publicized to help settlers provide for their crops and livestock—even where there was
no access to water. Since the 1870s, commercially manufactured
windmills had been used to drive reciprocating pumps that could
reach water to a maximum depth of 30 feet. But within decades,
the development of centrifugal pumps and drilling rigs powerful
enough to create large bore holes increased the volume and depth
of water available to farmers in areas previously thought too dry
for agriculture.

The main technological problem confronting agriculture on the
High Plains became how to supply the pump with a constant and
sufficient power source. Experiments with labor-intensive steam
engines and even electric motors eventually gave way to pumping
stations powered by natural gas or gasoline engines. Some of these
were simply modified Model T's, whose elevated rear axle drove
a belt attached to the rotor of a centrifugal pump. Gradually, volume and depth increased, but so did cost.[16] These problems were
not fully resolved until the late 1940s when engine parts and fuel
became widely available after the war. After that, irrigation on the

High Plains went into high gear, and drawing down the Ogallala Aquifer became serious agricultural business. The aquifer is now greatly depleted; the remaining water frequently comes from depths of 3,000 feet (over half a mile) or more. Water drawn from such depths is expensive, especially at a time when the average price of gas reaches $3.69 per gallon. Moreover, once the volume of an aquifer is depleted, concentrations of minerals like arsenic and cadmium make the remaining water much more dangerous and much less useful. Aquifer depletion, mineral poisoning, and rising water costs will soon end the era of agribusiness on the Plains. No one now knows where America's bread will come from in the 2030s. This issue became extremely pointed during the summer of 2012, when the unrelenting drought of an extreme La Niña event exacerbated by climate change destroyed 80% of the US corn crop.

The Awful Truth

Although no one on the Plains wanted to admit it, by the 1980s the problem of aquifer depletion and population loss was obvious. Far away in New Jersey, academics Frank and Deborah Popper were tracking the Western population decline county-by-county. The region had lost about 33% of its people in the 67 years between 1920 and 1987. The Dust Bowl ended in 1940, but the period of extreme outmigration that resulted from it was really just the beginning of a larger and much more prolonged migration. By the 1980s, about 3% of the country's population still lived on the Plains. When the Poppers began to map the ongoing decline, they realized that a huge north-south cross-section of the American West was returning to frontier. In fact, although the term "frontier" refers to areas with fewer than six people per square mile, in many of these areas there were fewer than two. Still, about six million Americans remained on the Plains in 1987.

That very year, the Poppers went public, predicting that the drying tendencies of the region would continue—for two reasons. Groundwater depletion, they pointed out, was becoming extreme in places like central Kansas where "in 1950...the Ogallala Aquifer

was 58 feet deep." It had shrunk to a depth of less than 6 feet in
the intervening years.[17] Moreover, as Deborah and Frank observed,
there was a powerful second reason to predict increasing aridity
throughout the West:

> The long-term outlook is frightening.... The greenhouse ef-
> fect—the build-up in the atmosphere of carbon dioxide from
> fossil-fuel combustion—is expected to warm the Plains by an
> average of two to three degrees making the region even more
> vulnerable to drought.[18]

The continuous aridity resulting from climate change will dehydrate soil throughout much of North America, making future
agriculture impossible in previously fecund places like America's
breadbasket. This is exactly what happened in the summer of 2012,
when America's corn and soy crop was almost entirely lost. But, as
early as 1987, the Poppers were able to predict that "massive swaths
of land would be abandoned as water for irrigation became unob-
tainable."[19] With the very real prospect of widespread depopula-
tion of the West, the Poppers earnestly suggested "de-privatizing"
139,000 square miles of extremely arid land, returning it to natural grassland, and making it into "Buffalo Commons," a nationally
protected bison preserve. When it was published, their study was
widely covered by the Western press. But the reaction they got from
Westerners surprised them. It was what you might expect if you
walked into a bank waving a gun.

It didn't really help matters that in the year immediately following their essay a drought gripped the region and stressed Westerners to the breaking point. Politicians, journalists and farmers took
turns vilifying the Poppers, ridiculing their ideas in print and in
public for the next decade. Hostility toward them was often extreme. Westerners reacted strongly to the suggestion that water was
going to become a much rarer commodity. Of course, this is not
such an unexpected response; limited water meant their livelihoods
were operating on borrowed time. When the Poppers' ideas were
first introduced, politicians were only slightly less caustic. In 1988,

Kansas Governor Mike Hayden responded sharply to the Poppers saying:

> America's Great Plains do not equal the Sahara. Why not seal off declining urban areas while we're at it, and preserve them as museums of 20th century architecture?[20]

The Poppers' unpopularity spread. As Frank dryly observed, he and Deborah scored "higher recognition ratings on the Plains than some sitting governors."[21] It was a very trying period for them. However, by the late 1990s it had become clear that the Poppers' predictions about the West's declining water and population were based on extremely thorough research and their analysis simply reflected the known facts. Today, there are only about 2 million residents left in the region, and it is widely accepted that there is considerably less water than when the Poppers first came to public attention. Outmigration, as the Poppers predicted, has left the West desolated, and the regional hydrologic decline they observed is ongoing. Early in the new century, communities like Clovis, New Mexico, predicted accurately that the aquifer on which their population depended would essentially be gone by 2010.[22] In addition, large portions of west Kansas, west Nebraska, and Oklahoma and Texas are already completely deserted. The number of ghost towns grows each year.[23] The ghost towns that fascinated Frank Popper have become so prevalent that a movement of outdoor travel fanatics called "ghost-towners" now routinely find, research and visit towns like Lake Valley, New Mexico, which was founded in 1878 and whose last resident left, at age 92, in 1994.[24]

Survival Techniques

These days, in the drying regions of the West, the last remaining citizen of a small town is usually an old person who has no special adaptive gifts but who has simply stayed on out of love and habit— until ongoing desertification overwhelms them. To some extent, before they leave, all Westerners try to imitate the survival strategies they observe in the well-adapted local life-forms: cacti, of course, conserve moisture for future use; desert toads and ephemeral plants like *ocotillos* evade the heat by resting or lying dormant until the arrival of rain; desert lizards and snakes have developed skins that reflect or redistribute heat while retaining water to resist dehydration. Humans have developed technologies to imitate each of these strategies, but more and more, aging Westerners are adopting the survival strategies of birds by simply escaping to cooler, more humid regions away from the unrelenting droughts and temperatures of the rapidly warming West.[25] Oddly, the word *ogallala*, which named a tribe of Sioux in the Dakotas, actually means "to scatter one's own people," so no matter whether you speak English or Lakota, flight, migration and exodus were already very old and acceptable survival strategies born of deep necessity long before climate change began to permanently dehydrate the High Plains.

In the warming West, Frank and Deborah Popper are now invited to speak at public forums that feature the same politicians who once criticized them. They accept speaker's fees, but over the years they have continued to refuse grant money for their unique and controversial research. Expressing the ethic of independent scholars everywhere, Frank explains why he refuses grant money: "We prefer to remain un-bought."[26] In 2004, former Kansas Governor Hayden (who now serves as Secretary of the Kansas Department of Wildlife and Parks) paid the Poppers handsomely to speak about the Buffalo Commons at a Kansas State University forum. During his introduction, Hayden ate a considerable portion of crow. "The truth is, I was wrong," Hayden said by way of an apology:

> The Poppers ended up being somewhat conservative in their estimates of the out migration from the Plains. The losses have actually exceeded most of the Poppers' predictions.[27]

Clearly, when they first proposed that the abandoned land of the Great Plains be returned to grassland and repopulated with herds of American bison, the Poppers touched a nerve among harried Westerners. Some understood the Buffalo Commons as a smug, unfeeling assertion by Easterners that the West was in an inescapable decline. Others felt that they were about to be evicted from their beloved homes. Nothing surprised the Poppers more than the negative reaction to their findings. Their idea was simply that since the American West was already emptying, the region should be given a second ecological chance. They have worked tirelessly without pay to "do right" by a portion of the American "biosphere [that is] in distress."[28]

In the fall of 1998, the Poppers' ideas achieved support from an unexpected source when the journal *Wild Earth* published a revolutionary essay by a pair of conservation biologists. Michael Soulé and Reed Noss extended the Poppers' idea of restoring wilderness areas by advocating a new a technique they called "rewilding." They criticized the management of America's national parks, which had allowed the unchecked loss of natural species to the combined challenges of foreign species, disease, and inbreeding, all of which were further complicated by the lack of natural predators. Soulé and Noss advocated "rewilding" these bio-geographic islands by identifying core species and reintroducing missing predators like wolves to select among them naturally. Their essay was but one sign of the expanding acceptance of the Poppers' dedicated advocacy of the Buffalo Commons and the philosophy surrounding it.[29]

The controversy surrounding "rewilding" has taken wild directions in the decade since the introduction of the idea, but already, wolves have been reintroduced to America's parklands, and their safety is guaranteed by court order. Reintroducing grazing animals to the abandoned lands of the High Plains seems a small step in comparison. Experiments in such restorations are already taking place. In the year 2000, Canada turned a tiny portion of southern Saskatchewan into Grasslands National Park—following the Popper model. This park which lies almost due north of Billings, Montana, includes the spot where Sitting Bull and 4,000 other Sioux crossed into Canada in 1876 a few months after the Battle of the Little Big Horn. On both sides of the border, the buffalo prairie still offers a flat, open, easy crossing. Sitting Bull couldn't have known when exactly he had crossed into what he called "the Grandmother's country," but these days there is a conspicuous blaze marking the park's southern boundary at the 49th parallel. As the subtropics continue to creep poleward, people may once again come this way as they leave the High Plains.

The Diminishing Cryosphere

Diminishing aquifers and soil dehydration caused by climate change have become decisive issues on the Plains, but many more Americans will be forced to deal with much less water—in the very near future. Shortages have already begun in the Southwest. When I first began this research, Governor Arnold Schwarzenegger had just ordered water from the Central Valley to be redirected to the thirsty cities of Southern California. It was the hot summer of 2009. Before the summer ended, wildfires and heat waves had exacerbated the state-wide water problem. Both California and Colorado were forced to tap agricultural water supplies in order to meet human water needs. Irrigation in Colorado declined 24% in the five-year period between 1997 and 2002.[30] Predictions from a variety of sources now forecast that agriculture across the Southwest will decline 30–35% in the coming decades as a result of global warming.[31] The new era of aridity and dryness will radically reduce the habitability of densely populated areas across the American West and Southwest. In addition, as the lack of water curtails agricultural production, it will simultaneously impair regional industries and have very negative effects on the national economy. Please believe me when I say, I am not happy about this. Nonetheless, it will happen.

Finding the Water

The future has already come to Los Angeles where so many people are from out-of-state that a kind of social seniority attaches to those who have the most earthquake stories. Even rarer than earthquakes are the snowstorms that occasionally occur within the city limits. Snow in the Los Angeles desert basin is probably the oddest phenomenon of life in this offbeat, desert megacity. But close-by in the mountains, blizzards can block the roads—as they did in the Tejon Pass during the winter of 2008. In the nearby San Gabriel Mountains, the snowpack builds during the winter months, so even on a burning hot day, Angeleños can drive their low-riders and BMWs less than an hour into the hills to refresh themselves on a "pista de esquí," or ski slope.

I've already mentioned that the cryosphere contains one half of all our earth's freshwater. Nowhere is its importance felt more powerfully than in Southern California. The snowpack is responsible for maintaining the year-round, water-rich lifestyle of the densely populated, semi-arid SoCal region. Most of the water travels 242 miles along the Colorado River Aqueduct from Lake Havasu, Arizona (where, curiously, the London Bridge is now located). The water carried by the aqueduct is mainly melted snow. Up north in San Francisco, where water rationing began in the spring of 2008, snowmelt also provides year-round water. But these supplies won't last. The IPCC (Intergovernmental Panel on Climate Change) estimates that with global warming, reduced precipitation, increased evapotranspiration, and snow that begins melting earlier every year, California will experience a 41% reduction in its current supply by 2100.[32] That date may seem far off, but the process has already begun. In 2008, California received only 67% of its average annual snowfall, and by early summer it was clear that state water levels had not been lower since the drought of '88. The reliability of the state's water supply has become a recurrent legal theme. Judgments in Riverside and San Luis Obispo Counties curtailed commercial and residential developments because of the lack of available water. These judgments, in turn, are discouraging nearby development in Kern County and elsewhere.[33]

With 24 million people, SoCal is less populated than the TriState metroplex on America's East Coast. Nonetheless, it contains the largest concentration of people in any *single* American state. Unfortunately, the challenge of providing sufficient water for all Californians and especially for those in the dense area south of the Tehachapi Mountains is overwhelming. Los Angeles has enforced city-wide cuts that brought per capita consumption down from 180 gallons per person per day in the late 1980s to 155 gallons in 2005. But while this is a remarkable achievement, it is also a frustrating one since during this period the city's total water consumption remained constant because the population grew by 750,000 people.[34] Clearly, without additional water resources, California must either shrink its demand *or* its population.

Uncontrolled population growth is an additional burden to water supplies already threatened by overconsumption and climate. With 18 million people spread over six counties, Los Angeles was once the fastest growing city in the country, but that distinction has now moved to urban centers east of the Rockies—where 75% of all available water arrives as snow on the Rocky Mountains.[35] People are leaving L.A. in record numbers.

In 1930, there were only about 11 million people in the entire American West; about half were Californians.[36] At that time, the Colorado River was an abundant source of freshwater for the whole region. Today, along its 1,300 mile course, the Colorado provides water for about 25 million people in California; another 10 million—in Arizona, Colorado, Nevada, New Mexico, and Utah—rely on the great river as their exclusive source of freshwater for drinking, agriculture, landscaping, sewage, and everyday use. The region has clearly outgrown its over-apportioned main source of water, and that source is shrinking rapidly.

And then there is an even bigger problem...

At an elevation of 10,000 feet, in "The Never Summer Wilderness" where the Colorado is born, spring has arrived for the first time since the last ice age. The snowpack has decreased sharply in recent years, and snowmelt now starts about four weeks earlier than it did in the 1970s. This means that by the middle of summer there are drought conditions in the largest cities of the Southwest. As a result of years of recent drought, Lake Mead and Lake Powell have each lost about one third of their volume. The perimeters of these depleted reservoirs resemble those disgusting rings that stain the interior of filthy bathtubs. Moreover, there's an increasing likelihood that these reservoirs will *never* recharge. Filling Lake Powell to its normal level of 3,700 feet would require above-average rainfall for a full five years—something extremely unlikely to occur for centuries.[37]

Like these reservoirs, the mighty Colorado has also been diminished by mismanagement, by a reduced snowpack, and by over-apportionment. These days the river no longer drains into the Sea of Cortez. Instead, it peters out just across the Mexican border. Writing about the beautiful river he visited with his father during childhood, Robert Kennedy Jr. lamented, "the Colorado River has nothing more to give…a train wreck is imminent."[38] Sadly, he was right. The IPCC predicts that in the coming decades the Southwest will become at least 30% drier. A declining snowpack and accelerated snowmelt are simply the first steps in reducing the carrying capacity of this vast region. An exodus seems inevitable. Similar changes are taking place throughout the continental United States. In Montana's Glacier National Park, there are now only 30 of the original 150 glaciers that were placed under federal protection in 1910. America's cryosphere is in the last stages of a ragged retreat.[39]

Back in Southern California, global warming's impact on the dynamics and volume of snowmelt has brought special problems to one city in particular. About 80% of the water supply of California's second largest city comes from local rivers fed by the snows of the Sierra Nevadas. San Diego's water source is already decreasing. In desperation, the city now purchases about 250 million tons of agricultural water from the Imperial Valley in order to supply human needs. In 2008, San Diego invested in desalination with considerable urgency, hoping at that time that as much as 15% of the city's water might eventually come from the ocean. There are 12 desalination projects planned in California for the coming decade; the first desalination plant, in Carlsbad, was designed to be the largest in the Western Hemisphere. It could supply 50,000,000 gallons of water daily, or about 9% of the region's total needs, at a start-up cost of $1 billion.[40]

The Carlsbad project is proving very costly. San Diego now expects to pay between $3 and $4 billion for Carlsbad's desalinated water, which will satisfy only 10% of the city's needs.

The reverse osmosis technique that Poseidon Resources uses to access the immense reserve of the Pacific Ocean requires a lot of expensive energy (used to force seawater through synthetic membranes that remove the mineral salts). As the energy-rich Kingdom of Saudi Arabia already knows, reverse osmosis is a quicker, cheaper and colder method of desalination than distillation (the only other option), but it still makes for very expensive water. Drawing both from the Red Sea and the Persian Gulf, Saudi Arabia now desalinates 743 million gallons of water a day and sends it far inland to supply cities like Riyadh and Taif. It is because of the volume and expense of Saudi Arabia's needs that desalination technology is developing quickly. An innovative process called electro-dialysis reversal (EDR) now complements reverse osmosis by cleaning the desalinating membrane-filter quickly and efficiently and bringing costs down. If additional innovations make desalination technologies cheaper, they could eventually solve California's water dilemma. For the next decade, San Diego will probably lead the way.[41]

Unfortunately, no kind of desalination will help the landlocked cities of the Southwest because they have no access to seawater. As a result, the era of expansion for America's fastest growing population centers is now ending. Phoenix, Las Vegas, and St. George, Utah all rely on the Colorado for water. Their unprecedented growth is now severely limited by the lack of water resources. Desertification has become a genuine and looming threat in America's driest states. Unfortunately, the water conservation technique most suited to helping Southwestern cities is also one that makes many Americans squeamish. *Wastewater recycling* meets with considerable intolerance wherever it is proposed. No matter how rigorous the process is, no one likes the idea of drinking and washing in purified sewage. Space Station astronauts may drink purified pee out of necessity, but on the earth's surface, there are very few volunteers.

For this reason, in 2001 Los Angeles scrapped a $55 million project to recycle wastewater that would have provided sufficient annual water for an additional 200,000 Angeleños. Oddly, Los Angeles found money enough to seed the clouds over the San Gabriels with silver oxide—in the unproven hope that it will produce 15% more runoff—but it will not use the much more reliable water-recycling technology. Similarly, in San Diego, Mayor Jerry Sanders opposed plans to launch a $6–8 million pilot project to recycle wastewater. Because of public prejudice, Sanders sees sewage purification as an unsellable political position, but his City Council disagrees. They voted against Sanders, and approved a pilot project in December 2007. This, in turn, became the prototype for a much larger $238 million water recycling plant. Council President Scott Peters summed up San Diego's desperation for water when the pilot project began: "We're just not in a position to turn up our nose at any option to increase the water supply."[42]

Even before drought was officially declared in 2008, Peters knew that his city would continue to face serious water shortages. Situated on the border with Mexico, San Diego also shares a climate zone border between the Mediterranean semi-arid climate of coastal California and the hot-steppe climate of Baja and Mexicali. In their desperate and growing need, many other Californian cities are now keeping close watch on how San Diego's decision to recycle plays out. Meanwhile, old preconceptions are being adjusted. As the state gets thirstier, Californians learn by example, and they're becoming more sophisticated water consumers. The world's largest wastewater recycling project opened recently in Orange County where the local population is expected to grow 16% before 2030. In January 2008, a new plant in Huntington Beach began producing 70 million gallons of treated water daily (enough for a half million people) at an annual cost of $29 million. Despite a price tag of $500 million, the ongoing drought made Orange County's wastewater processing plant look like a remarkable and sensible bargain. And in 2008, managers at the Huntington plant cleverly included a program of public education in their start-up costs which now seems to be paying off. Californians are changing their minds about drinking recycled water.[43]

There are good reasons to recycle wastewater. Recycling uses the same basic "reverse osmosis" technique as desalination. As these technologies develop, it may become possible to overlap recycling with desalination, making coastal cities much less vulnerable to water shortages. Reverse osmosis removes all the impurities—organic, mineral, pharmaceutical, microbial—from wastewater; and so it provides an end-product as pure as distilled water, purer in fact than the modern Colorado River which now contains a lot of the same human waste that makes Californians squeamish. Recycled water is probably much purer than anything most living Californians have drunk previously. From experience, I know it wouldn't take much to improve on the tea-colored liquid that used to come out of my tap in east Hollywood.

Megafires

Besides drinking water, there is a final problem that California experiences as a result of reduced snowmelt. This is something it shares with the Southwest, and it's a monster. Since 2003, scientists have understood that increasingly early springs result in a lengthening of the dry season, which in turn has a direct bearing on the size and frequency of wildfires throughout North America. Wildfires are usually caused by lightning or carelessness, so the devastating combination of summer winds and increasing dryness in the West have acted like steroids for forest fires. America has now adopted the word *megafire* to describe the overwhelming devastation of these infernos. The age of megafires likely began in 1988 with a blaze in Yellowstone Park that—despite the best efforts of 25,000 firefighters at a cost of $120 million—lasted throughout the summer until the first snowfall in mid-September; the megafire burned 600,000 hectares (1,500,000 acres) of forest.[44]

Since 1988, wildfires have gotten progressively worse; global warming is dissolving the snowpack earlier each year, making Western forests much drier much earlier. On average, forest fires used to last about 7½ days. Now they last an average of 37 days, and that average increases every year. Between 2000–2005, wildfires in the United States destroyed about 6 million acres of forest per year. Then, in 2005, wildfires took it up a notch, consuming 8.25 million acres of American forests—an area about the size of Pennsylvania. In 2006, this record was surpassed when another 8.7 million acres of forest burned.

The next year, megafires consumed 500,000 acres of Californian forest from Santa Barbara south to the Mexican border. The worst fires were centered around San Diego, and the two largest fires—the Witch (Creek) Fire and the Harris Fire—exceeded the size of the Cedar Fire, a raging inferno that destroyed 280,000 acres of forests, 2,800 homes and killed 15 people in 2004. Altogether, 500,000 Californians were evacuated as part of an immense firefighting effort that cost $20 million *a day*. Since 2008, earlier snowmelt, higher summer temperatures, and a longer fire season combine annually with the expansion of vulnerable areas of high-elevation forests to create a monstrously expensive problem that will continue to disrupt human lives and alter the habitability of western North America from Canada to Mexico for the rest of our century. From 1990 to 2000, 61% of all new housing in California was built inside or next to fire-prone forests, and, during the last decade, the amount of resources spent every year to combat California's megafires increased 150%—to over $1 billion per year by the summer of 2008.[45]

As the century progresses, and as California and America face an ever-increasing assortment of environmental challenges, money in these astronomical amounts will become increasingly harder to find. Undoubtedly, some people will consider outmigration as a reasonable response to the yearly devastation of the arid Southwest's forests and farmland. The process may have already begun in the Central Valley where, for generations, about 2 million Mexican migrants found sufficient fieldwork to purchase homes and settle into an American way of life. But by the fall of 2008, loss of agricultural work due to drought sent many Mexican citizens southward to share the burden of their unemployment among strong family networks in their native land. At the same time, the youngest and most mobile Latino American citizens in towns like Mendota and Firebaugh, California traveled to Alaska, Washington, North Carolina, and Canada to find work. Less than half of them are expected to return.[46]

Outmigration from California has not reached epic proportions. Recent years show a new total of outmigration between 150,000 and 250,000 people per year. But in Southern California, where warming conditions and fires are most severe, outmigration from urban centers like San Diego and Los Angeles is significantly higher. In these centers, the combination of economic and environmental collapse is now spiraling downward, just as it did in Mexico.

The changes I've described are gradual and cumulative. But nonetheless they have deep implications for the habitability of an enormous portion of the United States. Totally arid regions—like the depths of the Rub' al-Khali in Saudi Arabia—are genuine deserts, void of life. But even during extreme drought, the condition of "semi-aridity" permits life to continue. North America provides us with a beautiful example. In the deepest deserts of the Southwest, nature has blessed no other species with the perfectly adapted physical traits of the kangaroo rat, a truly amazing creature whose consummate gifts are sleekly tailored for life in semi-arid America. Unique among fauna, the kangaroo rat manufactures its own water by changing the molecular structure of its food through a biological miracle that rivals transubstantiation. As long as starchy foods— seeds, nuts, kernels—remain on the desert floor, these amazing creatures can persist. They do not sweat. They rarely pee. Their urine contains a much greater concentration of waste-to-water than is possible for mammals with less specialized and well-adapted kidneys. Since the early 20th century, biologists have marveled at this little rodent's "remarkable power" to exist "largely...on the water derived from air-dry starchy foods."[47]

Put very simply, kangaroo rats use the resources created by past water to gamble on the likelihood of future water. In fact, they have adapted so successfully to their semi-arid environment that they have an aversion to open water of any kind and even react badly to getting their noses wet. Unfortunately, not long after human beings leave, prolonged and repeated droughts and fires will turn the Southwest into a completely arid zone. Without the residual food of past years littering the desert floor, the region will become inhospitable even to kangaroo rats. This tiny miracle of adaptation cannot stand against the power that mankind has unleashed on its own environment. At night, when I look on the face of my sleeping son, I wonder if human beings will fare any better.

CHAPTER 4

Seaboard Diasporas

A loose-knit group of scientists and emergency managers was
the only thing standing between New Orleans and its eventual
fate.... They were small voices calling attention to ingrained
systemic problems.... Their heroic efforts ultimately failed....
It is too late now to save the New Orleans that once was. But it
may not be too late to do something about the storms to come.

MARK SCHLEIFSTEIN AND JOHN MCQUAID,
PATH OF DESTRUCTION (2006)

The United States loves its coasts. From Coney Island to Galveston, boardwalks and beaches are staples of American culture. It's possible that three centuries of ship-borne migration left traces that may be evident in the national devotion to the shoreline, but this much is certain: a majority of Americans choose to live within 50 miles of an ocean. For these people, "sittin' on the dock of the bay" names a national pastime not shared by nearly as many Mexicans, Canadians or Europeans. Americans, after all, were sailors long before they became cowboys. Whale song, waterways, Marines, and Navy Seals run silent and deep through the national psyche.

High Water

It's hard to guess how much of the country lives near saltwater, but nearly half the country shuddered collectively when Hurricane Katrina made landfall. By the time the storm ended, the significance

of unchecked climate change had finally come home. Despite puzzled looks on the faces of Federal Emergency Management Agency
(FEMA) Director Michael Brown and his president, in the early fall
of 2005 ordinary Americans began to realize that extreme storms
and rising sea levels might, as a senior English scientist would later
write, threaten the lives and homes of half the country:

> Whether or not Hurricane Katrina [could] be attributed
> directly to global warming, it certainly served as a wake-up
> call for how dramatically Nature can bite back even within a
> "safe" urban environment, and how foolish it can be to ignore
> the warning signals.[1]

Today America faces a threat more significant than terror, one that
will strike unrelentingly in many areas again and again. If the world
could somehow restrict carbon emissions to 450 ppm, the planet
would only warm about 2°C by 2035. But even this small change
will increase the intensity of rainfall, flooding, and the destructive
power of all storms. Obviously, these changes have already begun.
Unfortunately, it is now clear that the absolute lowest temperature
increase we can hope for is 4°C and that an increase of about 6°C
is extremely likely.[2]

With these temperatures, winter rainfall across the continental
United States will increase substantially. Warmer temperatures and
earlier springs will mean much less water will be stored as snowpack, so flooding will also increase significantly. Moreover, as the
planet heats up, thermal expansion will enlarge the volume of water
on the planet's surface and elevate current sea levels. Expansion currently accounts for an average annual sea level rise (SLR) of about
3–4 mm, but there is much more to come. Hotter sea surfaces mean
increased hurricane activity, so the 100-year-storms that reached as
far north as New York City in 1815 and again in 1938 will become
more frequent. Superstorm Sandy (which began as a hurricane)
was the third once-in-a-century storm to strike the northeast coast
in the second decade of our new century.

The IPCC's 2007 Fourth Assessment Report's projections for the rise in average world temperatures from climate change were based on the stationarity (regularly predictable increases) in world temperatures of the 20th century. But there is a problem in using the 20th century as a statistical baseline because most of it passed before the anomalies of rapid climate change became obvious. Still, the best or most conservative of these 2007 estimates put the ceiling for 21st-century temperature rise at 3.2°F. But, unfortunately, the most likely scenario (with a 66% probability) set the most probable rise at 7.2°F, with an upward ceiling at 11.5°F.[3] Today, widespread deglaciation is accelerating both in Greenland and the western Antarctic. The last time this happened (14–15,000 years ago), the oceans rose 20 meters (65 feet) over a period of about 400 years.

Things will happen much faster this time.

While the complete deglaciation of Greenland and the western Antarctic would contribute to a sea level rise of 7 meters (22 feet) and 5 meters (16 feet) respectively, the progress of ongoing partial SLR is really hard to calculate. The IPCC's overly cautious warning several years ago that a 0.5 meter rise could reasonably be expected in the next century didn't take into account the melting of the western Antarctic or the speed at which the world's ice sheets (ice on land) are now becoming seawater. The conversion of sea ice to seawater is now happening at about three times the predicted rate.

A Cambridge-trained member of the British Antarctic Survey stationed at Rothera recently observed a 7% increase in the ice-melt of the Pine Island Glacier (part of the western Antarctic Ice Sheet) which alone could raise sea levels by 25 cm (10 in)—or half the IPCC estimate for the coming century. A complete melt of the Pine Island Glacier in the near future seems increasingly possible and would result in a 1.5 meter rise.[4] Moreover, the melting of Arctic sea ice, which was previously thought *not* to contribute to sea level rise, apparently will contribute slightly. The world's oceans will rise an additional 2.5% of total melted sea-ice volume because of the differing densities of freshwater from ice and ordinary seawater.[5]

The Arctic ice sheet cracked definitively into three huge chunks in the spring of 2008; because these smaller pieces will melt faster, Arctic ice is predicted to disappear completely by 2030 (and probably much sooner).

What is especially significant is the discovery that surface water penetrates into the lowest layers of the world's ice sheets through open holes known as *moulins*. Underneath the ice, unfrozen water acts as a lubricant that accelerates the glacial movement of an icestream downhill from its position on the coast into the ocean where warmer water accelerates the melt. This movement drives the edge of the ice sheet into contact with an ocean at greater force and increases the whole process of global, or eustatic, sea level rise.[6] At Pine Island, a glacier 30-km wide is entering the ocean at the speedy rate of 3.5 kilometers per year (about 2.1 miles). One hundred square kilometers of ice makes a small but immediate difference in sea level, so by 2035, the sea level will most likely rise *several* meters (about 9–10 feet). If you live on a coast at sea level (as I do), a 0.5 m rise would just make the ground floor of your home unserviceable, but a 3 m rise would put a foot or so of water on the floor *upstairs*.

In the West, the elevation of the Rocky Mountains protects much of America's coasts from floods. Most vulnerable are the low-lying port areas like Los Angeles and Seattle, which are unfortunately also some of the most populous regions. The Pacific Institute finds that a 1.4 meter sea level rise stretching from Oregon to Mexico would put nearly 1.2 million people at risk of flooding and endanger $100 billion worth of seaside infrastructure, including 30 power plants, 28 wastewater treatment facilities, and both the San Francisco and Oakland airports.[7] On the Atlantic coast in the coming decades, rapid deglaciation will alter and erase much of the American seaboard—from the Greater New York region, southward. States on the Atlantic and Gulf Coastal Plains with low-lying expanses include New Jersey, Delaware, Maryland, the Carolinas, Georgia, Alabama, Florida, Louisiana, Mississippi and Texas. The shoreline of all of these (plus the District of Columbia) will relocate irretrievably during the next 30 years, and masses of people who live in these areas will either have to move peremptorily or be put into flight—unless significant and extremely costly preparations are made, including voluntary relocations and shoreline "hardening" along the 2,200 miles of the American mainland's Atlantic coast (more than a third of which—800 miles—is now considered a sea-level-rise "hotspot").

One in 13 Americans (about 22 million people) live around the hub of New York City, which has roughly 600 miles (960 km) of coastline spread throughout the five boroughs. In addition, the entire New York Metropolitan region has nearly 1,500 miles (2,400 km) of combined island and mainland coastline. This culturally and economically vital region is especially vulnerable to flooding from pretty much the whole variety of climate changes, including increased storm activity and sea level rise. But New York is only one of a number of vulnerable regions. Many mid-sized cities (including Charlottesville and Richmond, Virginia; Hartford, Connecticut; Harrisburg, Pennsylvania; Paterson, New Jersey; and Waterbury, Vermont) are clustered in river valleys far inland from the Atlantic coast, but will still be overwhelmingly affected, as will the Southeast, where a 3 m rise in sea level could potentially displace 8.7 million people, while a 5 m rise (the absolute maximum now anticipated for the coming century) could shift 17 million Americans.[8] The entire TriState coastal megacity that encompasses Greater New York has a population of about 60 million people.

In 2012, Asbury Sallenger at the US Geological Survey in St. Petersburg, Florida, published a study in *Nature Climate Change* identifying 600 miles (1,000 kilometers) of America's East Coast as a "hotspot" where sea level rise is now accelerating three to four times faster than along other US coasts.[9]

New Orleans lies outside of this hotspot, on the Gulf Coast. But because Katrina forced Americans to consider the vulnerability of all of its coasts, including the East Coast—the nation's demographic and cultural core—it has had a disastrous impact on the nation's self-confidence. There is no other way to explain why so much ink has been spilled rehashing the events—hundreds of books have been written since the hurricane took its deadly aim on New Orleans in 2005. Unfortunately, to date, this collective anxiety has not been addressed, nor has it resulted in any adequate preparations for the inevitability of coastal flooding right here in North America.

Katrina

The impact of a Category 5 (most severe) "Big One" hitting the rickety parishes of the Big Easy had been predicted with certainty since the computer modeling software called SLOSH (Sea, Land, and Overland Surge Heights) was first designed in 1978. In the 1970s, Category 5 storms were extremely rare, but maps based on the SLOSH simulations showed that the levee system could be overtopped by the size of sea wave (*storm surge*) that might precede even a Category 3 hurricane. Still, Hurricane Katrina was downgraded to a Category 3 storm by the time it made landfall. The fact that a storm measuring in the mid-range of the Saffir-Simpson scale could actually drown New Orleans raises a lot of questions about how we predict hurricane damage. Recent investigators note:

> At landfall…Katrina had weakened to SS3 status but the marine-equivalent wind field in the storm core expanded…. Katrina at landfall had destructive potential for storm surge similar to that attained a day earlier when it had stronger [more immediately intense winds]….
>
> In fact, Katrina had much more destructive potential than the SS5-rated Camille, which struck in 1969. So the tremendous wave and storm surge-related destruction of Hurricane Katrina at landfall was badly represented by the SS [Saffir-Simpson scale].[10]

It is a sad coincidence that Katrina came ashore just 23 days after an American researcher published definitive evidence that annual hurricane *intensity* was increasing as a result of the rising sea surface temperatures of global warming:

> The potential destructiveness of hurricanes…increased
> markedly since the mid-1970s…. Longer storm lifetimes and
> greater storm intensities…suggest that future warming may
> lead to an upward trend in tropical cyclone destructive po-
> tential, and—taking into account an increasing coastal popu-
> lation—a substantial increase in hurricane-related losses in
> the twenty first century.[11]

This account appeared online in *Nature* on July 31, 2005. Very few American politicians or bureaucrats knew anything about it when Katrina struck in late August. Moreover, the writing had already appeared on the levee walls by late March 2004. Freakishly, a storm with all the characteristics of a hurricane appeared far outside of the normal hurricane zone. Forming over the southern Atlantic, it cut a destructive swath through the Brazilian state of Santa Ca-tarina, becoming unofficially known as "Hurricane Catarina" for that reason. It had an amazing force. In Torres, where it made land-fall, it destroyed 30,000 houses. Near Brazil's Mampituba River, it lofted an entire house into another state; and at Passo de Torres, it wrecked an extensive shipyard in moments. Although the appear-ance of hurricanes in this region had been predicted by meteorolo-gists attentive to global warming, such occurrences were set for a much later date in the climate change regime. And while the occur-rence of extreme weather events before predicted dates is already tiresomely familiar, the range of this prediction is remarkable. Hur-ricanes were not expected in the South Atlantic until 2070, when it was thought sea surface temperatures would achieve a permanent level of about 30°C.

By the summer of 2005, the Caribbean had already reached that temperature, and surface temperature in the South Atlantic ex-ceeded the 26.5°C minimum that conventional wisdom offers as the basic condition for hurricanes. The extreme storms that followed included four Category 5s—more than any previous year. In July, Hurricane Dennis set the record for the most powerful Category 4 storm—with the strongest ever-recorded monthly pressure—before ripping into Cuba at full force. That same month, Hurricane Emily surpassed Dennis's record for pressure before setting another kind of record by becoming—briefly—the earliest recorded Category 5 storm over the Atlantic basin. After Katrina, there would be two more Category 5s, bringing the annual total to four. Wilma, the second Category 5 after Katrina, was the most intense storm ever recorded over the Atlantic. Wilma was the 13th hurricane of an amazingly bad hurricane season—the worst ever—and it caused $30 billion in damage.[12]

The extreme weather events of 2004 and 2005 were simply the last in the long chain of indications and warnings that preceded Katrina. In the late 1990s, FEMA listed a dozen possible disaster scenarios for the United States. After the #1 possibility—a terrorist attack on Manhattan—became reality in early September 2001, Scientific American decided to run a story about FEMA's #2 choice. Mark Fischetti's "Drowning New Orleans" appeared on October 1, 2001 and it did not equivocate. "New Orleans is a disaster waiting to happen," Fischetti wrote. "A direct hit is inevitable. Large hurricanes come close every year."[13]

Fischetti's piece is a riveting, excellent read. Nonetheless, it fell on the deaf ears of a nation numbed by 9/11. It was left to others to raise the alarm about an imminent disaster in Louisiana. At the Times-Picayune offices in New Orleans, Mark Schleifstein had been investigating local hurricane risks and pressing his editors to educate the people of the city about them since the 1990s. Despite the fact that he and writing partner John McQuaid had won a Pulitzer in 1997, Schleifstein had to undergo a rigorous process to gain editorial approval for a series about a dull environmental topic like levees and hurricanes. The hesitancy of the Times-Picayune editors is instructive for anyone writing to alert the public about unacknowledged environmental dangers. First, Schleifstein's editors were concerned that he would simply produce a collection of boring environmental pieces describing events that might not occur for decades. Then, they worried that by focusing on worst-case scenarios, Schleifstein's stories might be dismissed as sensationalist "disaster porn."

For these reasons, the *Times-Picayune* asked Schleifstein and McQuaid to focus primarily on the link between hurricanes and coastal erosion in the region and to fold the story of the impending "Big One" into this larger perspective.[14] This, they felt, would lend immediacy to the stories and enable them to do a much better job of getting the word out to New Orleanians.

As it turned out, they were right. Schleifstein and McQuaid's pieces changed the minds of many city residents about evacuating before the Big One hit. Although Katrina *was* responsible for about 4,200 dead or missing people (some estimates go as high as 11,000), organizations like the American Red Cross had been predicting for decades as many as 100,000 deaths if the levees were overtopped by a Category 5 storm surge.[15] The reason given for this huge number was the traditional New Orleanian indifference to hurricane warnings. Planners estimated that, as usual, only 60–75% of the city's residents could be relied on to flee the big storm when it came. This would leave about 200,000 people behind in New Orleans and, given the estimated 27-foot (8 m) storm surge of a direct hit by a Category 5 storm, about 30% of these were expected to die. "There's a reason New Orleans is famous for the drink named the hurricane," a Louisiana State University sociologist once quipped, "The culture here is 'We don't evacuate.'"[16]

But the *Times-Picayune* series, "Washing Away," seems to have muted some of the hurricane foolhardiness of Big Easy culture. As a Louisiana scientist would later write, the Katrina evacuation was "far more successful than anyone had imagined."[17] McQuaid and Schleifstein's articles instilled real fear in the hearts of New Orleanians. Stark descriptions—like the following—of the human costs of the coming Big One undoubtedly saved a lot of lives:

> Thousands will drown while trapped in homes or cars by rising water. Others will be washed away or crushed by debris.

Survivors will end up trapped on roofs, in buildings or on high ground surrounded by water, with no means of escape and little food or water.[18]

After the *Times-Picayune* series in the summer of 2002, New Orleanians became much more concerned about storm safety. Evacuation compliance for Hurricane Ivan in 2004 was higher than it was for George in 1998, when only one third of New Orleanians left town.[19] Before Katrina hit, there was a growing sense that New Orleans was past due, having, in the previous five years, dodged four bullets. By August 2005, the city had

experienced four hurricane-related evacuations:.... Dennis in 1999, Isidore and Lily in 2002 and Ivan in 2004. None of these warnings resulted in the level of destruction typically forecasted and popularly imagined.[20]

Quite likely, the best news of the whole Katrina story is the number of people who *did* leave in time: over 80% of New Orleans' citizens had left prior to the 29-inch surge that destroyed the levees. Unfortunately, there was no official effort to assist people in fleeing the city; most people who did not own cars were unable to respond to Mayor Nagin's late-blooming evacuation order. In the "antediluvian" New Orleans of 2005, there were a lot of people without cars. The city ranked fourth out of nearly 300 metropolitan areas whose residents lacked access to cars.[21] About 9% of New Orleans households didn't own or have access to any vehicle at all. Since about 53% of poor black people lacked access to cars, while many fewer poor white people (17%) lacked similar access, the majority of those left in the city to face the storm were black. Not surprisingly then, a majority of the storm's victims, those evacuated days after the storm, and those who were killed or went missing because of it, were also black.

Eighteen-year-old Jabar Gibson understood this inequality completely. On September 1st, he "commandeered" a school bus in Crescent City before picking up 70 friends, neighbors and strangers and driving 351 miles to Houston. The trip took seven hours and included three lengthy stops while he and the passengers scrounged gas and diaper money. Still, his group was "the first busload of New Orleans refugees to reach the... Astrodome." [22] Knowing where the keys were kept in his schoolyard, Jabar simply climbed a fence, broke a window, and left New Orleans in a big hurry shortly before noon with a handful of neighbors who had survived the floods. A newspaper account the next day reported "they took the matter into their own hands... because they felt rescuers and New Orleans authorities were too slow in offering help. 'They are not worried about us,' said Makivia Horton, 22, who [was] five months pregnant" at the time. [23]

Along the elevated sections of Highway 10 through Louisiana's Atchafalaya Swamp, Jabar stopped often to pick up more citizens of New Orleans and Baton Rouge who were fleeing their flooded homes on foot. His bus mainly contained children, but there were also mothers with young babies and senior citizens in their seventies and eighties, all of whom had been walking toward Texas for two days. They were often barefoot because their journey began with a cold, toxic swim away from their sodden homes. [24] When Jabar's group reached Houston at 10:30 that night, they were refused entry by Astrodome officials who said the facility was being held in reserve "for the... people being evacuated from the... Superdome." [25]

At the Superdome in Louisiana, Jack Colley, the governor's Emergency Management Director was overwhelmed when he discovered the original plan to accommodate 2,000 refugees was woefully inadequate. He phoned Judge Robert Eckels, the senior state official of Harris County, Texas, where the Astrodome is located, asking him to prepare 23,750 beds for all the flood refugees then at the Superdome. When Jabar Gibson drove up to the Astrodome, Eckels was busy inside. Outside, his officials quickly realized that Gibson's bus was an independent operation, not part of the Superdome evacuation for which they were completely unprepared. Not surprisingly, their default position was based on "lifeboat ethics," and they simply refused entry to Gibson and companions. An article from the *Houston Chronicle* from September 1st, 2005 is the only account of this incident as it took place, and it is valuable because it documents Gibson's disbelief and frustration firsthand:

> Now, we don't have nowhere to go.... We heard the Astrodome was open for people from New Orleans. We ain't ate right, we ain't slept right. They don't want to give us no help. They don't want to let us in.[26]

In the years since Katrina, Judge Eckels has spoken extensively about what transpired that night. There were huge tactical difficulties in sheltering nearly 30,000 Katrina refugees in the Astrodome. Eckels is quoted by historian Douglas Brinkley in his compendious history of Hurricane Katrina, *The Great Deluge*. Describing the exhausted young man he met outside the Astrodome that night, Eckels said:

> Gibson just jumped in a bus and drove.... He just grabbed a bus and drove it here, picked up people on the highway and drove them to Houston. That was the first bus. We had other people showing up. He got off the bus and people saw him and asked, "Where's the driver. We need to park this." And he was like, "What do you mean? I drove it."[27]

Generally, the predominantly black victims of Katrina experienced the same sluggish assistance that made young Jabar Gibson desperate enough to take matters into his own hands. Desperation comes quickly to flood victims and is actually a fairly sensible reaction—after any flood, urban centers quickly become toxic wastelands. Floods mix

> human waste together with chemicals from submerged kitchens and inundated industries, petro-fuels from service stations and pathogens shed by rotting corpses, both animal and human. When they retreat, they leave behind a thick scum of toxic muck, stagnant ponds attractive to mosquitoes,

waterlogged buildings and possessions that quickly become nurseries for mold. Thousands of houses survived Hurricane Katrina's initial assault only to be condemned…when fungi blossomed behind saturated walls.[28]

Many of the survivors of Katrina felt that America had left them to die in a toxic soup, and accusations of institutionalized racism have been circulating ever since. Undoubtedly, such racism existed. Media reports, for example, described Katrina victims in black parishes as hoodlums "looting" neighborhood stores, but similar words weren't used to describe white flood victims helping themselves to supplies in the abandoned stores of their neighborhoods. Of course, like Jabar Gibson, most people were simply trying to survive the impressive odds that Hurricane Katrina imposed on those left in New Orleans, but false rumors of black gangs wasting "white guys…they don't care which ones" were passed among white citizens, police, and reporters. Occasionally, they found their way into print.[29] Similar rumors were spread about the people lodged at the Superdome before the Astrodome was made ready. Judge Eckels's observations are relevant here. He thought rumors about black violence had to be exaggerated:

> All day we were getting ugly reports from the Superdome.… Governor Blanco called me and said, "There are criminal-type people; there have been a lot of problems at the Super-dome. You need to be ready for it." What we found was just the opposite. The people we got here were tired, they were in poor physical condition. They had been without food or water for several days, picked off the roofs of their houses or from a bridge. They had a long bus ride, some of them 12 hours or longer coming from New Orleans. But they were nice, courteous, no problem at all really.[30]

After Judge Eckels intervened, Jabar's group was accepted and given shelter at the Astrodome; many others were sheltered elsewhere in Houston and some in Little Rock. But local authorities routinely ran criminal record checks against their names and leaked spectacular cases to the local press. When they ran Jabar Gibson's name, they speculated about why Gibson knew exactly where the school bus keys were kept because he had a prior conviction for stealing automobiles. For a time, Houston police considered charging him, but eventually dropped the matter. No one has ever published an interview with Jabar Gibson himself about the events in New Orleans following Hurricane Katrina. While Brinkley was writing his book, Gibson was not available. He had been arrested in 2006 on unrelated charges and had been moved from Louisiana to the Federal Correctional Institute in Texarkana, Texas.

It took me three years to find Mr. Gibson's current whereabouts. I followed a false trail to Oakland, California, before I learned there were two Jabar Gibsons in the continental United States. Finally, I wrote him a letter in prison, and a few weeks later he called. Over the phone these days, Mr. Gibson sounds older than his 26 years, as well as considerably wiser and slightly tired. It was a faint connection, and his voice reminded me of the night I was lucky enough to catch Aaron Neville as he introduced the last tune of a very long set. Jabar Gibson told me

> Yeah, that was seven, eight years ago.... I'm still proud of what I done.... The people who started with me, they were my neighbors, y'know. When we left the city I knew 'em all. There were people on the road, too. Children wid' no shoes, old people, women with babies. It was a school bus. We had no gas but we had plenty o' room. We picked 'em all up.... Ran out o' gas a few times.... Had to beg for change.... I made a lotta bad decisions in my life since then, y'know. I'm payin' for them now. But that night...well, I know I did the right thing. I'm still glad fo' that night.[31]

As these accounts of the aftermath of Katrina suggest, a strong element of racism dogs the entire history of flooding in New Orleans. John Barry's *Rising Tide* is an eloquent history of how, in the past, flood events have become occasions for political intrigues and manipulations throughout Louisiana, either in order to retain a population that wants to leave or to drive undesirable elements away from the region. During the Mississippi flood of 1927, for example, a Delta planter tried to prevent his black sharecroppers from leaving the flooded region by imprisoning them in a cotton gin and nailing the doors and windows shut. At the same time, when other white planters in the area were rescued by a steamer, they tried to prevent over 200 blacks from boarding with them; they would have succeeded, except for a brave stand taken by a white American Red Cross physician. Even in 1927, not all white people fled, however. In flooded Greenville, Louisiana, 4,000 white residents lived in relative comfort on the upper stories of the town's remaining buildings, while about 18,000 blacks crowded into warehouses and mills or lived in the seamy deprivation of an elongated tent city that "snaked more than eight miles along the levee." During the crisis, the steamer "Capitol" supplied Greenville's white residents with whatever they needed, including newspapers on a daily basis, but then played "Bye, Bye Blackbird" on the ship's calliope every time it left.

The worst excess of the flood, however, was the deliberate destruction of levees in the outlying St. Bernard and Plaquemines parishes by the white, moneyed citizens of New Orleans. A group of leading New Orleanians successfully lobbied Louisiana's governor to dynamite portions of levees downriver from the city in order to create an emergency spillway that would relieve pressure on embattled levees upriver. They could just as easily have dynamited other levees in much less populated regions, but chose not to.[32]

Apocryphal or not, in 2005 this history made black community leaders genuinely attentive to depictions of the Katrina survivors. Quite soon, they began to focus on apparent biases in the media reports. During an early visit to Houston's Astrodome, Reverend Jesse Jackson criticized the media's choice of the word "refugee" saying, "It is racist to call American citizens refugees."[33] After an interval of several years, Reverend Jackson's claim can appear confusing since the word "refugee" carries no racial implications. William

Safire pointed this out at the time in the *New York Times*: "A refugee can be a person of any race at all." Safire wrote, "A refugee is a
person who seeks refuge."[34]

Even so, many African Americans felt that using the word "refugee" to describe the native-born evacuees of New Orleans robbed
them of their dignity by inviting comparisons with stateless people
like Wyclef Jean, leader of hip-hop group The Fugees, who had fled
Haiti to seek refuge in the United States. President Bush, however,
is not known as a hip-hop fan, so it was an odd moment of empa-
thy when the president attacked the use of "refugee" to describe
New Orleans's flood victims. Whatever his reasons were, Bush Jr.
said, "The people we're talking about are not refugees.... They are
Americans, and they need the help and love and compassion of our
fellow citizens."[35]

Undoubtedly, the president's intentions were good, but when
the beliefs that underlie his statement are unpacked, they seem to
explain the nearly total lack of preparation for Katrina and other
American environmental disasters brought on by global warming
or climate change. The president implied that genuine refugees *cannot be* Americans because refugees are people who come to America
seeking its good grace. Americans, quite simply are better—luck-
ier—more favored—than refugees. This is a peculiarly nationalistic
twist on the habit of blaming the victim described in Chapter 1.

The upside of believing that the USA is favored by fortune has
been America's confidence in the face of overwhelming challenges
like the naval war with Japan after the destruction of the American
fleet, or the nation's remarkable decision to visit earth's moon. Over
the centuries, America's reassuring conviction that "God is on our
side" has proved a valuable tool. Our brand of laissez-faire capitalism relies on a problematic tenet: If you trust in God, America
will prosper, and God will not do anything permanently bad to the
US of A. But until the belief that "it can't happen here" is rejected,
disaster preparations will continue to be treated with very little
seriousness in the United States. As long as America trusts com-

pletely in God and believes it has the inside track with Fortune, it will be very hard to convince people that extensive preparations are needed for anything.

Tellingly, naturalized Dutch scientist Ivor van Heerden relates an incident that took place during the Hurricane Pam simulation in 2004 when he suggested to FEMA that tent cities like those used to house the victims of war in Kosovo be prepared for an influx of New Orleanian evacuees. The response by FEMA was outright laughter and a curt rebuke: "Americans" a woman snapped in an acid reaction to van Heerden's mild-but-perceptible lowland Dutch accent "don't live in tents."[36]

It should be obvious that while such beliefs persist, America will never develop solidarity with the world's environmental refugees. It should also be said out loud that while this preconception lasts, environmental disasters will continue to be phenomenons that only happen "elsewhere" and to the less fortunate. This is a truly dangerous preconception, and, unfortunately, the coming decades will rob America of the luxury of such prefabricated "convenience-food" ideas.

Still, there is some encouragement even in these preconceptions. More than almost any other nation, America *is* privileged. Over-whelmingly, environmental disasters affect the poor more than they do any other group. So, for as long as the United States remains a wealthy nation, the environmental impact of climate change will not be as extreme as it will in a place like Bangladesh. But successive disasters will tax America's ability to protect its citizens over the long haul unless preparations begin very soon. Always devastatingly poor, Haiti is still on its knees after a series of Category 4 hurricanes (including Gustav and Hanna) destroyed island infrastructure and spread an aftermath of disease and violence in 2008. Extreme weather events are on the increase, and each one taxes our ability to afford the necessary repairs.

Serious and timely preparations, of course, are much cheaper in both human and fiscal terms than deaths, repairs, and restorations.

There is a brooding process of self-examination that follows disaster closely, like an unfed dog. In a state-sponsored study of the security risks that attend climate change, German experts made an observation that might be applied directly to the organizational and human disaster that followed Hurricane Katrina:

> In countries that lack warning systems or evacuation plans, extreme weather events cause relatively greater damage, and compel more people to flee than…countries that are institutionally well prepared for emergencies.[37]

Adapting to Change

The lesson of preparedness has already been absorbed in Bangladesh. In 1970, the Bhola cyclone (whose force was only equal to that of a Category 3 hurricane) killed as many as half a million people in what was then East Pakistan, one of the world's most densely populated and lowest-lying regions. By our standards, the average resident of modern Bangladesh is extremely poor, and before the lesson of these storm deaths was taken to heart, Bangladeshis did not have sufficient warning, resources or preparation to relocate quickly before the storm. In 1970, many of them died as a result.

Unlike the United States, Bangladeshis acted quickly to protect their fellow citizens. After Bhola tore up and scattered the country's infrastructure, Bangladeshis mobilized to restrict the human damage of future storms. Today, in order to spread the word of incipient floods and storms, cadres of whistle-blowing Paul Revere's fan out across the nation's roads. Despite a national literacy rate of only 41%, the noise of shrill whistles and bicycle bells in the streets moves the Bangladeshi citizenry immediately to higher ground. In 1970, cyclone Bhola convinced most Bangladeshis that God is best at helping those who help themselves. The human costs of remaining unprepared were revealed when Cyclone Nargis struck Bangladesh's neighbor Myanmar/Burma in May 2008, killing an estimated 140,000 people. And yet, the United States has done very little to prepare for a similar disaster.

In North America, where the spectrum of distributed wealth is far greater than it is in the subcontinent, the people who will be "immediately affected" and "affected most" by environmental disasters will be those with the least resources and the fewest options due to inadequate preparation. This is exactly what happened during Katrina. Besides the poor, these ready-made victims include children, the elderly, the infirm and the isolated. Such vulnerable groups must rely on the wealth and inclination of their governments to prepare for disaster and to see them through trying times. The double-shock of Katrina was that so little of America's wealth had been provided to prepare against an overwhelming and long-predicted disaster *and* that America's commitment to help its own victims was not strong enough to cut through a mire of bureaucracy. Although a Hurricane Katrina-style disaster had been war-gamed in 2004 as "Hurricane Pam," it resulted in no actual preparedness. Moreover, the Army Corps of Engineers responsible for protecting New Orleans from storm surges relied on an inadequate system of levees and pumps. It was known that a Katrina would eventually come, but no one did anything about it. The buck just kept being passed.

As every Oprah fan knows, women represent the greatest segment of the reading public, so perhaps the *Times-Picayune's* "Washing Away" series had its most powerful impact on literate female New Orleanians. During Katrina, it was mainly white female residents of New Orleans who *on their own initiative*, took their children out of harm's way well before the storm because they had access to information and the freedom allowed by financial resources. For this reason, white women were able to follow their instincts and leave the city early, even though during disasters, evacuation frequently means family separation. Women and children sometimes head off early for safer locales while men remain behind to resolve financial or residential issues. During Hurricane Katrina men—especially employed men—were far more likely to send their families to safer ground and then remain behind to guard their jobs and to protect the family home. Mark Schleifstein, for example, would have liked his wife to leave the city while he remained behind to cover the storm. But after their son agreed to take his dad's Prius to Atlanta to stay with his married sister, Mark's wife Diane felt her responsibilities toward her adult children had been discharged, and she simply refused to leave her husband's side. With formidable courage, she stayed in the city with him throughout the entire ordeal.[38]

In general, personal initiatives to move to a safer zone prior to anticipated disasters are much more common among women with children than among men, and they are most common among the educated middle and upper classes—those who can afford to spend money on disaster preparation "just to be safe." In New Orleans, fewer black women appear to have made the same choice as their white counterparts simply because there were many fewer black *middle-class* women in New Orleans. Like the issue of access to cars, this is just one more example of how, all too often, the survival choices of America's poorest citizens are bitterly constrained by their lack of resources.

The evacuees who left New Orleans after the storm were only a tiny fraction of those who left the city forever *before* Katrina struck. About half a million people left New Orleans by road in August 2005. Many fewer returned. Oddly, no one ever refers to "voluntary relocators" (those who jump before disaster strikes) as refugees. But these early responders are by far the largest number of migrants fleeing a given disaster. Although large, the cluster of later refugees usually represents a small fraction of the size of this initial movement. Altogether, Katrina set about 750,000 people in motion within days of its landfall. This is an early indication of the potential size of migrations due to climatic events, and if the terms "environmental refugee" or "climate refugee" are to have any meaning, they should also describe those people with sufficient foresight, resources and anxiety to relocate voluntarily when crisis looms. It is important to remember that without the return of the half million or so New Orleanians who didn't wait to see the floodwater before leaving, the city remains a pale shadow of its former self.

Worldwide, about 58% of all refugees (more than 25 million people) are victims of some kind of climate or environmental change. The United Nations estimated there would be 50 million climate refugees by 2010, while at Oxford, Norman Myers guesses that during the next 50 years, climate change will increase the number of environmental refugees to 150 million, and the IPCC has also suggested 150 million such refugees will exist by 2050. Obviously, estimates vary wildly and are confused by disagreements about what constitutes an environmental refugee and what kinds of climate change should be included. But strangely, those who leave before disaster overtakes them are never counted among official refugee figures.

Dr. Graeme Pearman, an Australian climate scientist, predicts that by 2100, a 2°C rise in temperature will place 100 million people worldwide "directly at risk from coastal flooding." We already know that this was a very conservative estimate since temperatures are expected to rise to between 2.5 and 3.5°C. Moreover, the Greenland ice-melt is occurring rapidly. Some projections place sea level rise as a result of the Greenland ice-melt at "6 feet this century."[39] But this seems radically conservative in comparison to other predictions that show 70% of California's less-dense snowpack disappearing by 2050.[40] The density of Greenland's ancient ice cap is supposed to slow ice-melt (as opposed to the melting snowpack in lower latitudes), but we now know there are special conditions that accelerate melting of icepack in the Arctic. While very little is yet published, the World Bank now acknowledges a much more rapid rate (three times faster) of sea level rise than anyone previously anticipated. Environmental statistic omnivore, David Suzuki, describes sea level rise in the next few decades with the phrase "several meters," and this seems to be the consensus among all the scientists and environmentalists I have interviewed. To my knowledge, only the World Bank is a bit more specific:

> Continued growth of greenhouse gas emissions and associated global warming could well promote sea level rise (SLR)

of 1 m–3 m in this century, and unexpectedly rapid breakup of the Greenland and West Antarctic ice sheets might produce a 5 m SLR.[41]

As we will see later, 2012 proved a pivotal year in our reassessment of the speed of Artic deglaciation; that summer, for the first time, nearly 97% of Greenland's entire ice sheet showed signs of accelerated ice-melt. We do not know how quickly Greenland ice will disappear into the sea, but it is already happening much more quickly than previously estimated.

How Bad Could It Get?

Three to five meters (10–16 feet) of rising water is extremely bad news for people clustered along the hotspot of America's Atlantic coast. Unfortunately, New Orleans was simply the lowest hanging fruit for the intensified environmental threats of global warming. Many other communities, including America's most concentrated urban center, will be increasingly susceptible to storm activity and sea level rise in the coming years. That's why environmental defense has become a powerful issue along the Atlantic seaboard. New Yorkers are, sadly, all too aware that much of their city's infrastructure is well below sea level. They're reminded every summer when the subways close due to flooding in lower Manhattan. But before Superstorm Sandy, very few people were aware that New York is vulnerable to once-a-century storms. Seventy years had passed since the hurricane of 1938, and each year brought increasing sophistication in our ability to plot and predict hurricanes. Still, as Katrina demonstrated, our ability to predict disaster accurately means very little if no one pays attention to such predictions.

The destruction of the Great Hurricane of 1938 was proportionately as terrible as that of Superstorm Sandy. At a time when infrastructure and population was much less dense, the hurricane tore apart Long Island, New Jersey, Connecticut, and Rhode Island, leaving 632 dead and creating the modern equivalent of $4.7 billion in damages. Over 20,000 buildings were demolished, and 75,000 were damaged, while 26,000 cars were ruined beyond repair. Sixteen hundred cattle died and, strangely, 750,000 free-range chickens simply disappeared because modern methods of "high density floor confinement" had not yet been developed by poultry farmers. The '38 hurricane was described as a 100-year event, but under global warming, such events will become increasingly more common, happening every 50, and then 30 and then 20 years apart. The wreckage left in the wake of these storms will also become increasingly expensive. The single bit of good news is that fewer people will die in each storm as we learn how to prepare for such assaults. Nonetheless, more and more hurricanes will form in the North Atlantic, and directly threaten the New England coast. A brave Vancouver woman brought this home in an account of her rowing trip across the Atlantic. In *Rowboat in a Hurricane*, Julie Angus writes:

> The formation of a hurricane in the cool waters above the Canary Islands seemed impossible.... The temperature of these waters was only 23 to 24 degrees Celsius; hurricanes are generally thought to need surface temperatures of at least 26.5 degrees.... [But] a storm with the strength of a ten-megaton nuclear bomb exploding was moving implacably toward us across the open ocean. I felt like a prisoner in a penitentiary about to be consumed by flames.[42]

Whether such storms come from the equatorial or northern Atlantic, America cannot afford regularly repeated disasters. The '38 hurricane accelerated the Great Depression in New England, destroying the tobacco, fishing, and textile industries and reducing property values to nil. It was estimated that the storm produced—in seven hours—enough felled lumber to build 200,000 five-bedroom homes. Long before FEMA became famous for world-class ineptitude, President Roosevelt sent 100,000 workers from the Works Progress Administration (WPA) and various other agencies, including the Coast Guard and the army, to set New England back on its feet.[43]

In the next century, global warming will accelerate the occurrence of 100-year storms like the '38 hurricane, while heat waves also become more intense and more frequent. And the same climatic change will bring a period of accelerated SLR, but how much is still unknown; the predicted range is disconcertingly wide, extending from the IPCC's extremely conservative 0.5 m to the more reasonable (but still conservative) level of 1–3 m predicted by the World Bank, to the extreme maximum of 12 m, which is what Kevin Trenbreth says would be the case if the earth's entire cryosphere melts in the next 100 years.

But, as Superstorm Sandy demonstrated, New York is not Atlantis, and it does not need a 12 m sea level rise to bring about disaster. At the end of 1992, the storm surge from a nor'easter was enough to disable metropolitan subways and turn many New York roads into lakes. In New Jersey and on Long Island, residents were evacuated.[44]

A 1999 study by the Environmental Defense Fund points out that necessary "adaptation may be costly and still leave the [Greater New York] region as a whole a less desirable place to live, visit or do business." Planning ahead should become a routine practice for any agency that might be affected by SLR in the New York region, since increases in sea level are inevitable. The alternative is simply to do nothing "until the need for action becomes obvious [since] it is notoriously difficult to affect change until disaster strikes."[45]

Janine Bloomfield, author of the study, makes a variety of suggestions, including building a sea-level wall around lower Manhattan. The expense of these preparations guarantee they will not be realized, but the only alternative is to abandon the Greater New York region to the fate suffered by Burma/Myanmar's Irrawaddy Delta when Cyclone Nargis struck in 2008. In 2005, Dr. Bloomfield summarized her findings about the dangers of sea level rise to New York, writing:

> The metropolitan East Coast Region is highly vulnerable to global warming and resulting increases in sea level. New York City has close to 600 miles of coastline. Four out of the

five...boroughs are situated on islands.... Entry points to many of the tunnels and much of New York City's subway system lie less than or near to 10 feet above sea level as do the three major airports.... Potential impacts of...sea level rise include widespread coastal and inland flooding...increased threat from mosquito-borne diseases, stressed water supplies and loss of beaches.[46]

To that cheerful bit of news, you can add the following observations by Mark Lynas:

Flooding problems [in greater NY] will be worsened by coastal erosion as rising sea levels gradually eat away at the shoreline. Without massive sand replacement exercises, beaches in New Jersey and Long Island could be moving inland by up to 3 metres a year by the 2080s.[47]

Neither Bloomfield nor Lynas explore the truly unmentionable fact: there are 6,000 miles of New York sewers, and all of them lie well under sea level.[48] Without preparations, a very small sea surge could turn the entire region into a toxic soup that would be ripe for cholera, which would bring the cultural capital of the world to a complete and abrupt halt. Asbury Sallenger, of the US Geological Survey, has suggested that there are sound reasons for such alarm. In the past, experts believed that sea level would rise proportionate to only two forces: the thermal expansion of water already in the ocean, and the addition of more water to that volume. But in 2012, months before the superstorm called Hurricane Sandy finally convinced many New Yorkers that climate change was happening, Sallenger realized:

Climate warming does not force sea-level rise (SLR) at the same rate everywhere. Rather, there are spatial variations of SLR...forced by dynamic processes...from circulation and variations in temperature...salinity, and...static equilibrium processes arising from mass redistributions.[49]

In particular, along America's northeast coast, as the Gulf Stream's northern extension slows, the slope of water in the ocean will change, lowering itself in response to the Gulf current as it peters out. This in turn will push up the level of the sea along the northeast coast *independently* of ocean surface temperature or the precise volume of Greenland ice entering the ocean as meltwater in the higher latitudes. In the coming decades, "anyone living along America's East Coast north of Cape Hatteras will see a more rapid increase in coastal erosion, sea surges and destruction of coastline infrastructure than in any other location in the continental United States. This destruction will only increase as the century progresses."[50]

CHAPTER 5

Urban Heat

*Heat waves sometimes leave us with an exceedingly
narrow margin of safety even here in... America.
When heatstroke begins to appear in the hospital, a rise of
another 5°F... would produce a holocaust of deaths.*

CLARENCE ALONZO MILLS, CLIMATE MAKES THE MAN (1942)

Some people believe that modernity and improvements in our technology will protect humanity from extreme climate conditions in the 21st century. For example, it is often suggested that improvements in air conditioning technology will likely completely eliminate the minor problem of urban heat waves.

Unfortunately, this is utter bullshit.

As cities increase in size, the problem of generating and retaining more heat becomes exponentially worse. Already during the intense heat waves of the new century, air conditioners require more power than urban electricity grids can supply. There are interesting experiments with new systems like "deep water cooling," but these are still very cumbersome and costly. It is safe to say that as heat waves intensify under climate change, their invisible lethality will increase also. Moreover, the problem of extreme heat waves is greater by far than the simple fact that air conditioners will tax electricity grids beyond their maximum capacity. By the second week of August 2010, over 700 Muscovites were dying daily from a deadly combination of a heat wave's extreme urban temperatures and the megafires ignited in the boreal forest surrounding Moscow where the canopy and undergrowth became as dehydrated as tinder after a few days of extreme heat. As the city's residents fled, AFP reported that Russian economic losses for the entire period of the extreme weather event might exceed $15 billion.[1] The loss of this money, of course, leaves all Russia increasingly vulnerable to similar climate events, which will be happening with greater and greater frequency.

Fire has always been one of the major feedback problems attending urban heat waves. If a city is plagued by intense *dry* heat, its cheapest, oldest buildings become increasingly susceptible to fire and present a perfect opportunity to arsonists. And, because inferior materials dehydrate more quickly than costly ones, once a fire begins, its severity is exacerbated by poor design. This became a vital issue in Manhattan years before building codes began to specify fire retardation standards:

> The central airshaft in many tenements [allowed] even the smallest fire to spread quickly between floors, a fire intended to destroy only a business or personal belongings could very easily consume an entire building and its inhabitants…."My youngest child is only six weeks old," Mrs. John Lyons told a reporter, "but she has already passed through two fires."[2]

In addition to these secondary complications, there is general ignorance about heat waves and death by hyperthermia. Few people know that heat waves are the *greatest* environmental killer. Because their impact worsens as cities increase in size and density, their power to cause large numbers of urban deaths is even greater in the 21st century than it was over 100 years ago—despite the fact that tenements have been outlawed. It may seem unlikely in 2013 to claim that heat waves will cause an exodus from the enlarged cities of mid-century North America, but intense heat lasting through the long months of summer will become a contributing factor to a prolonged urban exodus that may have already begun. Climate change, after all, does not happen alone. As the 21st century becomes hotter, America's infrastructure will decay. Water will become scarcer, and water suited for human use will all but disappear in major urban centers, where repairs to sewers, drains and purification facilities are already unaffordable.

The failure of roadways, urban services, and a sufficient, timely, and well-preserved food supply will also contribute to making life in any major urban center unendurable for many residents. In Texas during the 2011 heat wave, 10% of the state's shade trees died, making the state's urban crucibles even hotter. Wildfires then took 500 million more trees out of Texas's forests. But, as the urban trees died, they sucked every drop of moisture out of nearby pavements, producing cracks and gouges even as the heat simultaneously broke up concrete foundations of Texan homes.[3]

Nonetheless, the biggest danger of heat waves is still the human mortality that occurs through ignorance. In 2010, a record number of American infants died of hyperthermia because they were left asleep in cars during the summer heat wave.[4] Infants comprise one of the largest groups of heat wave deaths. A famous case from the 1896 heat wave indicates this is a constant concern. Mrs. William Grimm left her baby in his carriage in direct sunlight on W. 42 St. as she went inside to get her husband from their ground floor apartment. In direct sunlight, the temperature that day reached 135 degrees. An upstairs neighbor, pharmacist David Wheeler, noticed the child was in distress when he returned home. He took it to his apartment, stripped it and was bathing it in cool water when the distressed mother arrived. Fortunately, Mr. Wheeler had told a neighbor his intentions as he brought the child indoors. But still, he was arrested for kidnapping and spent a very hot night in a New York holding cell before a local judge dismissed the case the following day. The *New York Herald* ran a follow-up to the kidnap story with the headline: "Samaritan After All."[5]

Heat Waves Kill

If you believe in technological progress, 1896 was just a step or two from the doorway into our modern age. That year saw the introduction of dial telephones and x-ray photographs. Still, the country's agricultural origins remained obvious in the fetid litter that turned

American pavements into unhealthy thoroughfares. Four years before the dawn of the 20th century, only 25 automobiles had been
built in the United States. When exceptional temperatures burned
the streets in late summer, 19th-century Americans were torn between staying inside or wandering abroad with pocket handker-
chiefs pressed to their faces against the mephitic stink of rapidly
decaying horse manure. The dilemma was captured by *Scribner's
Magazine*:

> When the heat came and the sun on the pavements began
> to look white...the breath from the streets was such that no
> one knew which was worse, the hot, foul air outside, or the
> close, foul air inside.[6]

In some areas of New York, conditions were especially gritty during the dog days of summer: "The huddled and packed crowds on
the east side...live in the street, sleep on the roofs, and endure the
heat in silence," wrote a young member of one of New York's founding families in a memoir that contains few other criticisms of his
beloved home.[7]

The Lower East Side was ablaze during the heat wave of 1896, so
Colonel George Waring opened his heart and the city's fire hydrants
to the huddled masses. As New York's newly appointed Commissioner of Street Cleaning, he ordered the streets flushed daily in
an effort to sanitize and cool the city.[8] But despite the success of
Colonel Waring's efforts, the nearly 7,000 pounds of horse manure
and 160 gallons of horse urine that littered New York's streets daily
left the reek of decay throughout the overheated metropolis:

> The effect of...temperature upon the refuse and filth of
> the streets, courts and alleys...the air in close places, in the
> tenement houses, and upon the tenants themselves is soon
> perceptible.... Foul gases of decomposition fill the atmo-
> sphere...and render the air of...unventilated places stifling;
> while languor, depression and debility falls upon the popula-
> tion like a widespread epidemic.[9]

Still, not everyone gagged and wilted in the summer heat. Newspapers describe "heat refugees" who left for the seashore or the mountains, and Cornelius Vanderbilt II enjoyed a scrubbed and air-conditioned atmosphere at his townhouse on West 57th Street. In 1893, air conditioning pioneer Alfred Wolff installed this first residential cooling system at the Vanderbilt home after having created an ice-cooling system for Carnegie Hall in 1889.

The 1890s were a decade of extreme summer heat in New York, and, as the leading designer of a technology so new it had not yet been named "air conditioning," Wolff's star rose. In 1899, he constructed a cooling system to prevent cadavers from decaying at the 1st Ave. campus of Cornell Medical College. The system worked so well that, on hot days, school administrators propped the dissecting room door open to provide "comfort cooling" for the rest of the college. Before his death in 1909, Wolff successfully created cooling systems for the homes of Andrew Carnegie and J. J. Astor, as well as commercial systems for the New York Stock Exchange (1902) and the College of the City of New York (1905).[10] Cornell has remained a leader in cooling technologies for architecture and uses nearby Lake Cayuga for its "Lake Source Cooling System."

But at the turn of the century, outside the small world of Cornell's corpses and Wolff's rich clientele, summer conditions simmered. In the country, cows stopped producing milk and chickens died in swarms—as they always do during extreme heat events. Within the city limits, draft horses were among those affected most. A contemporary study noted that for horses,

> sunstroke is manifested suddenly. The animal stops, drops its head, begins to stagger and soon falls to the ground unconscious.... Breathing is [noisy]...pulse is very slow and irregular, cold sweats break out...and the animal often dies without recovering consciousness.[11]

By August 9th, 1896, horses were collapsing in the crowded streets of all eastern metropolises. Some animals simply died in harness, and many died unnecessarily since horses were often put out of their misery whenever it became obvious that volunteer bucket-brigades were too busy or too far away. Tragically, sometimes the horses' stressed and overheated handlers also died from the physical exertion of trying to move their mounts because it was widely believed that when a horse collapsed, its death was imminent. A paper described the perfunctory treatment of horses with heat stroke by inexperienced handlers:

> A pail of water dashed over the head, a couple of kicks to see how much life was really left in them—this was about the best a sunstruck horse got. And then a bullet to put him out of his misery.[12]

In Manhattan during the first exceptionally hot week of 1896, the Second Avenue Horse-Car Company lost 100 of its best animals.

It is difficult to quantify the sheer misery of a heat wave in turn-of-the-century New York. Fortunately, it was the golden age of magazines, and many of the era's most eloquent voices tracked the noteworthy events of the day capturing contemporary sentiments about the minutiae of life:

> A hot wave was telegraphed from the west, and the week… was an anguish of burning days and breathless nights which fused all regrets and reluctances in the hope of escape.[13]

Although ice was available throughout the 19th century as an industrial material, it became a widespread consumer commodity after 1861 when the "ice-box" experienced an enthusiastic debut. Before the ice-box, cold beer had been a treat reserved for delivery days when railroad refrigerator cars arrived in small towns throughout America. It was during the first summer of the Civil War that "ice-cold beer" became a permanent national staple. In 1896, there was no summer "ice famine" because the previous winter had been very cold. To keep their beer frosty, New York's saloon-keepers had plenty on hand. When the heat wave struck, they generously provided blocks of it to pack around their patrons' dying animals.

Their generosity was costly because Charles Morse, an ice-magnate from Maine controlled commercial ice production in Manhattan and kept the price of ice artificially high. After the heat wave, the daily paper in Morse's hometown would describe him as "the man who made millions while the poor people suffered for ice." The same article claimed lack of ice increased deaths among the poor by 5% due to Morse's chilling tactics.[14] If, in the early days of the heat wave, there was free ice for dying horses, there was none for New York's poor until Theodore Roosevelt, the new New York Board of Police Commissioner, effectively broke Morse's ice-trust by recommending an innovative policy with an unusual level of social responsibility for the 1890s. At a time when there was no social safety net in the United States, Roosevelt recommended the city itself take over from the barkeepers by providing the poor of the city with free ice. By touring the precincts, Roosevelt personally supervised the effort in order to prevent the usual rampant local fraud; the city provided $5,000 to distribute 95,000 pounds of free ice to its poor and infirm.[15]

Heat Islands

Apart from cooling human beings and animals who had collapsed outdoors, little else could be done to ameliorate the cumulative effects of heat on New York mammals. The "heat-island effect" permits no relief—even at night—so dispatching removal crews with block-and-tackle to hoist sick horses or their carcasses onto large wagons meant risking the lives of even more horses and men simply to keep the roads clear. Since there was really no cooler place to house the afflicted horses, street crews worked slowly, and the bodies accumulated.

Urban "heat islands" are a complex phenomenon. Amid the coolness of the surrounding countryside, large cities exude heat like lighted candles. For this reason, heat islands are sometimes called *inverse*, or *reverse*, oases. On the hottest days, cities accumulate heat and remain much hotter around the clock than their rural surroundings. During the still, clear days of a heat wave, man-made building materials absorb solar radiation from dawn to dusk. Until about 1997, all American roofs were built using asphalt or other dark, dense materials that have a very high heat-conductivity and storage capacity. And in large American cities, roofs comprise about one quarter of all urban land cover.[16] Because roofing materials diffuse heat very slowly, the amount of heat retained in urban settings is remarkable:

> Urban surfaces such as roofs and pavements are routinely heated...to temperatures 27–50°C [50–90°F] hotter than the air. Air temperatures in a typical mid-latitude US city range from 15–38°C [60–100°F] on summer days, and urban surfaces can reach peak temperatures of 43–88°C [110–190°F].[17]

In contrast, rural areas are substantially cooler. In the country, there is less pollution to trap heat; there are no urban canyons to prevent cooling breezes; and the vegetation of country settings itself is a natural coolant since evapotranspiration works constantly to lower the temperature of green areas.

But in the city, buildings are designed to preclude the presence of water, so very little cooling evaporation takes place. Moreover, the height of buildings, their closeness to each other, and the configuration of their roofs trap all the solar heat that enters the city all day long. This mixes with anthropogenic heat sources in the city—cars, stoves, fridges, machines, human bodies—and elevates the temperature of any urban area without vegetative cover.

At night, all this accumulated heat diffuses from the pavement and the buildings. Simultaneously, this process of diffusion warms up the canopy, the city's lowest layer of atmosphere. So, during heat waves, because of the heat-island effect, there is no relief during a nighttime period of cooling, and this is especially true in the densest regions of the inner city. In fact, a plume of hot, still air stretches like an invisible flame above the city's canopy into the "boundary" layer of the atmosphere. In 1978, the existence of this plume was discovered when a satellite equipped with infrared cameras studied weather conditions over Buffalo, New York.

If we were equipped with night-vision goggles at street level on a hot evening, we might see the heat plume above the skyscrapers in our own city's downtown core, and it would resemble a red-orange minaret or a flame. Together, this minaret and the skyscrapers below give the impression that the overheated downtown core is a massive torch or candle.

On the hottest nights, the difference between the open-air city temperatures and those of the surrounding countryside can reach 3.6°C (6.5°F). So on a very warm night when it is 94°F in the open air of the country, it is over 100°F inside the city.[18]

If this difference seems small or strikes you as a rare occurrence, it's important to remember that indoors, where the heat has been accumulating all day, nighttime temperatures are much higher. It is especially hard to know what interior temperatures were during 19th-century heat waves, since observers of the time were eager to take "objective" temperature readings on top of the tallest buildings so that they would be "free from local variations." Fortunately, in 1896, *Scientific American* noticed an anomaly between these objective readings and the blistering temperatures New Yorkers were experiencing:

> Heat in the streets of the city and in its stores and offices has often risen many degrees higher than the official records, and a street temperature of from 97 to 103 degrees has been common on such days as the 8th, 11th, and 12th.[19]

For ten days that summer, street-level temperatures were so extreme that horses and men dropped in the thoroughfares of America's largest and best cities. Municipal rescue attempts focused mainly on the human beings, so the figures for equine deaths are much higher than for human beings. *Scientific American* reported 1,500 dead horses, so it's safe to say there were at least that many.[20] But a barge captain who transported Manhattan's dead, rotting animals to Barren Island, remembers shifting 1,258 dead horses in the final four days of the heat wave.[21] Unfortunately there are no comparable figures for the number of dogs shot after they went mad, which often happens in extreme heat. Still, if you compare 1,258 horses over four days to the total human deaths in the entire metropolitan area for about ten days of excessive temperatures, the equine deaths roughly equal the excess human deaths—1,281—that are now calculated for the 1896 heat wave.[22] This figure makes the heat wave a worse natural disaster than the 1863 draft riots, the Great Chicago Fire of 1871, and the more recent Chicago Heat Wave.

Human Fatalities

It's hard to believe, but heat waves cause more deaths annually than any other natural phenomenon. When the 1896 heat wave was over, the *New York Times* estimated deaths in the city at 594 people. *Scientific American* placed the number at 632. But these numbers are probably large underestimates. As one medical researcher has wryly observed, "It is the exception rather than the rule for the medical effects of heat waves to be reported with precision."[23] Moreover, as events in Chicago during the summer of 1995 demonstrate, there are often official reasons for preferring low death estimates. The real death figure for the 1896 heat wave was more than double the number estimated.[24] In those days, Americans most vulnerable to fatal heat stroke were predominantly Irish outdoor laborers, including teamsters and hackneymen, longshoremen, hod-carriers, window washers, blacksmiths and the even the city workers of Central Park.[25] Many of them died from "exertional" heat strokes, just as the horses had died when their body temperature zoomed past the breaking point as they labored in the unusual heat of the day.

Cholera Infantum and Other Heat-Related Deaths

At the same time, there were many other kinds of deaths indirectly related to the soaring and souring temperatures. Urban children died from "the summer complaint," an annually recurring gastrointestinal disorder that proved fatal within days. This illness, which doctors of the era called *cholera infantum* was later determined to be a bacteriological infection derived from food, water or milk, which spoiled quickly in an era of poor refrigeration.[26]

There were also heat-related deaths among apartment dwellers in the tenements on the Lower East Side, where crowded, poorly constructed and badly ventilated buildings offered little relief, even at night. By 1895, the number of tenements on the Lower East Side had grown to 40,000, and there were about 1.3 million residents; density is estimated to have been between 250,000 and 300,000 people per square mile. By the turn of the century, 1.5 million people were jammed into 42,700 tenements in lower Manhattan.[27] So many people were crowded into these buildings that, as Arnold Bennett, a visiting British novelist, wrote "the architecture seemed to sweat humanity."[28] Theodore Roosevelt's description after a fact-finding trip to the tenements when he was a young assemblyman in the 1880s is much more descriptive and detailed:

> There were one, two or three room apartments, and the work went on day and night in the eating, living and sleeping rooms—sometimes in one room. I have always remembered one room in which two families were living. On my inquiry as to who the third male was, I was told that he was a boarder with one of the families. There were several children, three men and two women in this room...tobacco [for cigar making] was stowed about everywhere, alongside the foul bedding, and in a corner there were scraps of food. The men, women and children in this room worked by day and far into the evening and they slept and ate there.[29]

Unfortunately, Roosevelt does not describe the ambient temperature of these tenements in the summer months, let alone during heat waves. Daytime solar radiation penetrated deep into the fabric of these structures. Ventilation was so poor and the buildings were so close together that heat did not diffuse at night. Exhausted and hot, the occupants often sought respite by sleeping on rooftops which they called "Tar Beach."[30] They also perched precariously on window sills since even the still night air was much cooler than temperatures inside these urban ovens. Years before air conditioning became affordable and commonplace, many tenement residents fell asleep and then to their deaths every time the mercury soared.[31] Between August 5 and 13, 1896, at least 420 tenement dwellers died from a variety of heat wave-related causes, including falling to their deaths.[32] John Hughes was one of these. On the night of August 8, 1896, Hughes was so hot he couldn't sleep. He drank heavily, purchasing several "growlers" (chilled buckets of beer) from the local saloon and then drank them on the rooftop of his E. 98th St. tenement. As Hughes stretched out near the roof's low parapet, a neighbor in a nearby window warned Hughes that he might roll off in his sleep and be killed. Hughes ignored her. Later, he fell five stories onto the concrete pavement of his building's courtyard. His twisted, bloody corpse was covered with laundry from a clothesline that he encountered on the way down.[33]

Why was it so hot that summer?

Since the end of the Civil War there had been speculation that American summers were becoming hotter and more severe. In 1878, *Harper's* magazine put it this way: "There are those who have supposed...our seasons are changing, and...we are passing into a more tropical condition."[34] This speculation seems impossibly premature, since the genuine temperature increases of anthropogenic global warming would not be documented for another hundred years. But the past is full of insights. In the summer of 1896, an astute (and unfortunately anonymous) scientist commenting on the global nature of the heat wave wrote:

> It is probably more than a coincidence that heat waves of unprecedented power and duration should have visited...three continents...in the same year.... Science has yet to discover the influences which determine their coming and going.[35]

1936: Technological Improvements?

We now know there is a direct relationship between these "hot waves" as they were then called, and the La Niña phenomenon.[36] We also know that until global warming increased summer temperatures in North America, such massive heat waves were generational events occurring for a cluster of years and then stopping for more than a decade or so. There were bad heat waves again in 1911 and 1917 before La Niña conditions relented. The cycle restarted in 1932, and 1936 was the hottest year of this next cluster of truly exceptional summers. During the 1935 heat wave, California papers advertised vacations to Canada for $79.[37] But by that time, air-conditioned public spaces were numerous, and, except for police horses, most of Manhattan's hay-burners had been replaced by cars. Still, during the Depression, horses were common throughout rural America, and these animals just kept dying in the heat. In an account of his road trip from Cincinnati to Kansas during the southern heat wave of 1934, Clarence Mills, an intellectual pioneer deeply concerned with climate's influence on culture, wrote:

> Passing through the Indiana and Illinois harvest fields… horses were dying of heatstroke by the hundreds, and tales of human prostration greeted me at every place I stopped.[38]

Despite this description, there is no mention of large numbers of *urban* horse fatalities in any 20th-century source. So by 1936, technology appears to have triumphed over one of the challenges of exceptional heat in America's cities. In fact, technological improvements had softened the blow of many heat wave conditions (providing some justification for the belief that technological progress might protect us from the extremities of future climate change). For example, after WWI, public water purification achieved more modern standards. At the same time, better refrigeration, transportation, and public health criteria meant that urban supplies of food and milk were no longer regularly tainted by bacteria fatal to infants. Electric clothes irons had become common consumer items, so clothes were no longer pressed with irons heated over blazing indoor fires. And gas and electric stoves also produced much less waste heat in the domestic environment, as did early electric conveniences like toasters, electric kettles, and even much more specialized appliances like electric popcorn makers or sandwich irons.

The real innovation, however, was the electric fan, a localized cooling device that replaced industrial ceiling fans driven by belts attached to waterwheels. These were introduced as early as the 1860s. But in 1882, Philip Diehl put the powerful electric motor he had designed for Singer sewing machines into a ceiling fan, giving it the ability to circulate air quickly through multiple rooms. This development was followed in 1886 by Schuyler Wheeler's invention of the portable stand-alone electric fan. A generation of American children bore scars from these early devices because, at first, they had no protective cages. But by the 1930s, the caged, electric fan was a mature, affordable and ubiquitous product. It was also remarkably cheap when compared to the $1,000 price tag of the earliest consumer air conditioners.

The disadvantage of these fans was that they didn't cool, but only circulated indoor air. Still, during the 1930s heat waves, innovative city residents learned how to reduce the temperature of a single small room by setting an electric fan on a table and blowing its exhaust over pans of block-ice or ice chips. Competition among fan manufacturers was stiff during the Depression, and because all fans functioned with more or less the same efficiency, manufacturers competed with each other using appealing designs, and fans became very stylish consumer items. Emerson Electric Company produced my personal favorite, "The Silver Swan," a small, black and silver art deco fan now worth about $600 on eBay. In the extremely hot summer of 1936, every manufacturer of these electric devices reported an enormous upsurge in sales.

To keep cool, urban residents would also loiter in movie theaters and department stores, both of which had been air-conditioned since the 1920s. In the stores, they purchased record numbers of electric fans; but during the heat wave of 1936, American stores were completely out of stock. Sales of men's undershirts declined radically that summer, but people raided department stores for sun hats, light summer clothes and sun glasses (which were polarized for the first time in 1936). Americans also drank more soft drinks and ice-cold beer in cans that year than ever before, and the price of lemons (the raw material for lemonade and Tom Collinses, a popular extra-large cocktail of the era) jumped $2 a box.

Does Technology Help?

At first, technological progress (in the form of fans, ice, and air conditioning) appears to have helped people cope with the worst conditions of urban heat waves. But did it? More people were comfortable, it's true. But despite all their efforts to stay cool, people also died in greater numbers during the '36 heat wave than ever before. Why? Because cities had expanded rapidly since 1896. As they expanded, the heat-island effect trapped more solar and anthropogenic energy. So, as cities became larger, taller and denser, they also became hotter, and the heat lasted longer.

During the summers, temperatures in America's inner cities soared. In the blazing heat of 1936, bridges across the Harlem River expanded so much that they jammed open. In Detroit, by July 14th, there was a death every ten minutes, and the local hospitals and city morgue were soon completely filled. Moreover, there was no clear understanding of how to treat *all* heat wave cases effectively since "heat stroke" is really not one phenomenon but a confusing cluster of physiologic reactions to extreme heat. Causes listed on death certificates during heat waves often reflect these confusing conditions. Hyperthermic mortality can be called by many terms: *asthenia, convulsions, exhaustion, heat stroke, pyrexia, hyperpyrexia, insolation, sunstroke,* or *thermic fever.*[39] Sometimes, the ice-baths used to rapidly reduce the body temperature of those who collapsed from heat worked too well, sending overheated patients into a fatal shock. Ironically, hospitals were not air-conditioned by that time, so heat wave victims often died in stifling wards while waiting for treatment.

Modern emergency rooms aim to reduce heat stroke victims' core body temperature to 38.9°C (102.2°F) within 30 minutes of their arrival by immersing them in icy or very cold water. Body cooling units have now been developed to help doctors and nurses achieve these goals, but in the old days, success was a rare thing. During a 1917 heat wave in Chicago only one third of those treated survived; things had improved slightly by 1936.[40]

In 1933, medical researchers at Harvard had shown that severe summer heat had a cumulative effect on the human body after a period of three days. A statistical study done at the School for Public Health compared death rates during successive Boston hot spells. For the first time, doctors noticed a spike in heat wave deaths on the fourth day of the 1911 and 1917 heat waves; they also noticed a much lower mortality rate during the 1925 wave, which lasted only three days.[41] Despite the fact that it was a technical, medical report, the three-day mortality threshold for *hyper*thermia became accepted wisdom in the 1930s. Around the same time, a new category of heat wave victims was recognized: those who died by drowning. Heat wave sufferers made enormous efforts to cool themselves, and municipalities facilitated this by opening more and more public swimming pools. But at ten cents a swim, the price was sometimes too high for Depression-era kids. In Camden, New Jersey, 16-year-old Matthew Petrelli was brought into court for assaulting James Katoulos, 17, whom Petrelli accused of a very petty theft: "I wouldn'ta smacked him," Petrelli told the judge, "only I needed the dime to go swimmin.'" With the wisdom of Solomon, Joseph Lieberman, the presiding judge, collected a dime from those present in his courtroom before sending both boys to the pool under police escort.

In fact, swimming was the majority solution for beating the heat. On the hottest days of the 1936 wave, 600,000 people crowded the beaches of Coney Island (about 20,000 of these would sleep there every night for the next week). About 400,000 more swimmers crowded the beaches of Far Rockaway. Elsewhere around the five boroughs, people jumped off bridges or rented row boats from which they dove and swam. Each day saw more drowning deaths. These occurred when bathers misjudged the currents or their own level of exhaustion or experienced shock from cooling too quickly after entering the water. The highest number of deaths I found is 16 drowned on a single, simmering summer day. The usual daily numbers were lower, but sadly constant: three one day, five the next, ten on another day. On the impassive screen of a silent microfiche reader, all of these deaths still seem tragically wasteful, but a few truly stand out: on July 11th a 6-year-old boy drowned in the Harlem Mere of Central Park while on the same day, a 65-year-old woman, driven mad from the heat's effect on her high blood pressure, first threatened to commit suicide and was later found dead in Liberty Park Lake near her home.

1995: Chicago Burning

If it's inaccurate to include deaths by drowning or by falling from rooftops in the official count of deaths-by-heat, these deaths should at least remind us that heat waves have disastrous human costs. Numerically, they are the worst environmental killer of human beings, claiming many more lives every year than tornadoes, hurricanes, earthquakes, floods, winter storms or *hypo*thermia combined. Heat stroke kills about 1,000 Americans in an ordinary year. During heat waves, this number quickly multiplies to as many as 10,000 deaths. Death rates are highest in large, overheated cities. Widespread ignorance about the impacts of heat on the human body—especially in hot urban environments—seems to include a bias against seeing high summer temperatures as a severe mortal threat. This heat wave bias, or blind spot, became an especially pressing issue during the Chicago Heat Wave of 1995, which killed nearly 800 people and became the subject of a detailed study by sociologist Eric Klinenberg. His highly readable and well-researched account of this extreme event shows us how knowledge and technology can combine to protect citizens against climate change.

The decade following Richard M. Daley's election in 1989, marked a concerted effort by Chicago's administration to reverse the negative effects of decades of de-industrialization. Their common enemy was a prolonged period of localized economic recession described as:

thirty years of economic and social decline triggered by the flight of [Chicago's] manufacturing industries and the degradation of its neighborhoods and streets.[42]

By the summer of 1995, Chicago's recovery was in full swing, but then too, the first noticeable climatic changes began to affect continental America.

By previous standards, the heat wave that struck the city in the summer of 1995 was a monster. But it was hardly the equal of the extreme wave that would strike Europe in 2003; that one would last for weeks on end and would kill between 30,000 and 50,000 people. During the coming century, such events will become commonplace in mid-latitudes around the world, and in the United States, temperatures will reach into the 106–120°F range for weeks at a time.[43]

By 1995, Chicago's mayor and civic administrators had worked for six years to reverse the nation's negative image of their city. Two horrifying books, *The Truly Disadvantaged* and *When Work Disappears*, document the emergence of Chicago's black underclass and the city's deteriorating economic and social spiral following the disappearance of industry, prosperity and decent jobs. Despite the brash, autocratic style that is probably his most obvious inheritance, Richard M. Daley was determined to restore the city to the prosperity it had enjoyed under his father, Richard J. Daly. By slashing public programs and expenditures, he was having considerable success. Before the heat wave struck in 1995, Chicago had informally adopted as its motto: "The City that Works." But the number of heat-wave fatalities would soon threaten the positive image that Mayor Daley's motto conveyed.

Whether Daley knew it or not, Chicago was fortunate to have as Chief Medical Examiner, Edmund Donoghue, an uncompromising and highly trained pathologist who had worked in the examiner's office for 18 long years. Donoghue knew from long experience that heat waves are "slow, silent…invisible killers whose direct impact on health is difficult to determine."[44] To help his investigators, Donoghue established clear criteria to identify heat-related deaths. This was something neither the federal government nor the National Association of Medical Examiners had yet done, so Donoghue was an innovator—bringing a heightened level of insight to a very difficult job. Moreover, his criteria were straightforward and extremely practical. In the summer of 1995, a death in the city of Chicago would be judged "heat-related" if:

+ body temperature measured 105°F soon after death; or if
+ a body was found in a room without a/c, with closed windows or with a high ambient temperature; or if
+ a body last seen alive during the heat wave was found decomposed without no other visible cause of death.[45]

On Friday, July 14th, Donoghue alerted the city to its heat-wave crisis by publicly reporting 199 confirmed heat-related deaths. Because he knew the mechanics of cause-of-death reporting, Donoghue was aware that many deaths in Chicago would go unidentified and suburban heat-related deaths would not be tallied until a much later date. Mayor Daley reacted to Donoghue's announcement of these deaths by saying: "It's hot. It's very hot, but let's not blow it out of proportion."[46] Columnist Mark Royko of the *Chicago Tribune* picked up the mayor's theme and turned it into a front-page editorial whose headline directly challenged Donoghue's integrity: "Killer Heat Wave or Media Event?" it screamed.

Other members of the local press felt the mayor's response was a bit insensitive in light of so many dead people, and they pounced. While the storm gathered, Donoghue cautiously reviewed his figures and his criteria. He knew that the outcry was not going to blow over; uncounted corpses were piling up, and refrigerator trucks were being rented to accommodate the stream of bodies that had already filled the city morgue.[47] Meanwhile, because of the incredible demand that air conditioners placed on the power grid, outages were occurring throughout Chicago. Because of this, there would be even more deaths. In poorer neighborhoods, police and city workers were closing fire hydrants that citizens had opened to cool themselves and the streets. This did little to encourage good will. But water pressure was greatly reduced throughout

Chicago. In some areas people could not flush their toilets and only a trickle came out of their taps. Everyone—*the press, city officials and the citizens*—were hot, irritable, thirsty, and more than a little scared.

Donoghue had issued his public statement fully expecting shock and some criticism. He felt it was his responsibility to warn people. Even so, he never expected criticism from the highest levels of city administration. But when it came, he was unwilling to do the politically expedient thing and keep his mouth shut. This stand would eventually earn him the presidencies of the Chicago Medical Society, the National Association of Medical Examiners, and the American Academy of Forensic Sciences.

To be fair, Richard Daley did not immediately call Donoghue's figures into question. Still, the high number of heat-wave deaths threatened the image of the municipal makeover he had been struggling to achieve. Daley knew that Chicago once had an emergency heat-wave plan. But in an odd foreshadowing of the circumstances surrounding Hurricane Katrina, the costly plan had never been implemented. As Chicago's administration struggled to revitalize the country's third-largest city using tough policies of fiscal minimalism, they had slashed services to stretch the municipal budget and speed the recovery. While these policies were working, no one wanted to go out on a limb and spend the big bucks needed to prepare for a meteorological event and besides, it was only a heat wave, they felt, not something truly serious like a hurricane. This led the city to pursue a path of denial and recrimination before Mayor Daley woke up to the disastrous potential of heat waves and instituted a rapid reform.

Blind Spot

Heat waves don't visibly wreck infrastructure as extreme storms do, so their role as killers is constantly underestimated. Lethal heat stroke persists partly due to the widely shared disbelief that people don't die from something as simple as high temperature. This blind spot is an ongoing problem because the physical processes of heat

prostration are difficult to diagnose. Until you are actually overcome by heat, the danger of your situation and your proximity to
exhaustion may be unknown to you. This might be the reason that
so few people take heat exposure seriously. During my research,
I was surprised to find that at the beginning of each heat wave, news
agencies invariably apply humorous angles to stories about excessive
temperatures. A *New York Times* article from the earliest days of
the 1936 heat wave collects funny anecdotes about the impacts of
extreme temperatures across the Big Apple and bears the headline
"Heat Wave Has Its Comic Ripples."[48] Meteorologist Paul Douglas
recalls a frustrating argument he had with a TV executive at the
beginning of Chicago's killer wave. Douglas recalled talking to his
producers:

> We said, "This is going to be a major story. People are go-
> ing to be dying. This is something you'd better hit very, very
> hard…. I'll never forget [the executive producer]…wanted
> to do a live shot with some place…hotter than Chicago. She
> kept wanting…a featury, lifestyle kind of cutesy…story….
> I kept pleading with her…." You're missing the point. We
> should have people at the hospitals, we should have people
> at City Hall." It degenerated into a shouting match…. She
> started screaming "You don't get it! This is television!"….
> I said, "I do get it. I understand. This is a dangerous situa-
> tion for Chicago. We're the hottest spot. People will be dying
> later today. That's your story."[49]

In failing to enact their emergency provisions for heat waves, Chicago's administrators participated in the common inability to see
heat as a serious human threat. Historian Edward Kohn offers this
explanation:

> It is in the nature of heat waves to kill slowly, with no physi-
> cal manifestation, no property damage, and no single cata-
> strophic event that marks them as a disaster. For that reason
> [a] heat wave is only infrequently remembered.[50]

In 1995, Chicago's city officials didn't act until the crisis was upon them. Then, when people started dying, Mayor Daley undoubtedly felt threatened; 16 years earlier, his predecessor, Michael Bilandic, had lost his office to Jane Byrne over his mishandling of a different environmental disaster—a series of freak blizzards. So, the level of fear was palpable among Daley's administrators. An anonymous city official later recalled:

> The heat wave was the only time during my tenure when we were having closed meetings with locked doors and signs saying "No Press Allowed."[51]

But while hundreds of dead citizens might have been enough to hang a lesser mayor, Daley, a marine reservist, was made of much sterner, more practical stuff. He was also fast on his feet and adapted quickly. On the 17th, he met with city administrators to decide how to handle the crisis. The mayor's deepest concern was that "soaring mortality rates would create a public impression that his city was unprepared for the situation."[52]

Daley's team quickly scheduled two public relations events for the same day. One was an appearance at a senior's home, intended to educate citizens about how to stay alive during the heat. There was also an appearance at a north-side supermarket where power outages had shut down refrigerators and freezers and ruined the neighborhood food supply. In an attempt to shift responsibility, the mayor confessed to his own blind spot about the heat wave before going on the attack:

> Let's be realistic. No one realized that deaths of that high an occurrence would take place.... But people are angry. They're frustrated and want to go back to normal living. My office has been in touch with Commonwealth Edison throughout the power failure, and I'm not happy with their response. This is not over.[53]

Daley promised to conduct public hearings about the power outages. Then Chicago's Police Superintendent and various commissioners took turns defending the city's response to the press. All was going well until Daniel Alvarez, the Human Services Commissioner, blamed many of the victims of heat for their own deaths:

> We're talking about people who die because they neglect themselves. We did everything possible. But some people didn't even want to open their doors to us.[54]

Unlike Mayor Daley's canny attempt to shift responsibility onto a faceless corporate entity (already widely disliked due to a long history of slipshod electrical service), Alvarez's tactics excited a visceral reaction from families of the deceased and their communities, many of whom were black. A Chicago Public Health Department study would soon confirm that most of the heat-wave deaths had occurred among society's most vulnerable members: the poor, the elderly and African Americans, but the press had already reported this in daily stories.

And the worst was yet to come.

On July 18th, the recorded death rate from the heat wave soared to 376. And two *Chicago Sun Times* reporters had used their City Hall contacts to unearth the fact that Chicago had not implemented its own heat-wave emergency plan.[55]

In desperation, Commissioner Alvarez appeared before the press to apologize for blaming the victims of the heat. Later, Mayor Daley himself spoke to the press. It was at this point that he began to explicitly question Edmund Donoghue's figures:

> Every day people die of natural causes. You cannot claim that everyone who has died in the last eight or nine days dies of heat. Then everybody in the summer that dies will die of heat.[56]

With the scent of municipal blood tingling in their hairy nostrils, Chicago's veteran journalists interviewed Donoghue for his reaction. Although he was diplomatic, he absolutely contradicted the mayor, who, Donoghue said

is entitled to raise questions. We would be delighted to have
our figures checked.... But all of these [deceased] people
would have survived except for the heat. I think my criteria
are fair. I think my criteria are excellent. The truth of the
matter is we have probably underestimated the number of
deaths.[57]

Eric Klinenberg's dramatic book, *Heat Wave*, includes excerpts
from anonymous interviews with insiders about the crisis. According to a high ranking—but still unidentified—city official, after
Donoghue's statement contradicted Richard Daley, "the mayor
was nuts with Donoghue. He wanted him to shut up."[58] Local
TV news programs were now running headlines like "Who's to
Blame?"[59] On the quiet, Chicago's Public Health Department began an epidemiological investigation of Donoghue's figures. What
they quickly discovered, however, was sobering. Using the quick
and easy criteria of "excess deaths," which compares historical statistics of recorded deaths during a given period to current reports
of deaths, the Health Department found that 739 people (200
more than Donoghue reported) died in Chicago during the July
14–20 period.[60] Suddenly Mayor Daley accepted his coroner's figures, and publicly backtracked. He was no longer "questioning any-
body dying." On July 20th, he held a press conference and revealed
the city's brand "new" heat emergency plan (a hasty reworking of
the old one).[61]

This might seem like a cynical reaction, but it's really just business as usual. Mankind does not willingly prepare for crises. Crises
must first overwhelm us, humiliate us, and bring us to the brink of
defeat before we rally and mount an effective counterattack. For
this reason, I believe little will be done and many people will have
to die in the coming decades before we take the threat of climatic
change seriously enough to mobilize collectively. A recent British
book about climatic change describes the potential danger to mankind's continuing existence:

Heat waves of undreamt-of ferocity will scorch the Earth's surface as the climate becomes hotter than anything humans have experienced before throughout their whole evolutionary history.[62]

In 1995, the mayor of Chicago successfully sidestepped the heatwave crisis and then implemented a new plan of action for heat waves. Later that summer, when the heat wave returned, the city acted quickly to locate and warn its most isolated and vulnerable residents and move them to the assortment of cooling centers set up throughout the city. Proportionately, there were many fewer deaths during that second wave, as well as the one that followed in 1998. Preparedness is a key issue in confronting extreme weather events—but preparedness is expensive.

Richard Daley learned from his mistakes and educated himself extensively about heat islands and heat waves. The year after his close call during the heat wave of 1995, Daley toured green roof sites in Germany and then began a municipal policy unmatched by any other American city for more than a decade. Chicago now experiments with and encourages green roofs throughout the greater Chicago region; this is a good thing because they are effective at reducing temperature in heat islands. The positive city image that Mayor Daley was eager to create prior to the 1995 heat wave was given substance through a painful learning process that began during the crisis. As a leading landscape architect has written: "Chicago used green roofs as part of its strategy to project a revitalized green image."[63]

Chicago City Hall itself is a prototype for a new technology that could help other American cities survive the coming era of extreme heat waves when summer temperatures will range between 106 and 120°F for weeks at a time.[64] America will need this new technology as the destructive force of heat waves grows. In the coming century we should expect severe heat for all US megacities on the East Coast. In Los Angeles, heat waves will last longer, quadruple in frequency, and produce temperatures that hover around 116°F.[65]

Not everyone is convinced this is a serious problem. Two MIT economists have confidently predicted that (except for infants) few Americans will die of excess heat during the coming century (provided energy costs remain roughly the same *and* we spend $15–35 billion per year to power our air conditioners).[66]

But, as Mark Lynas has pointed out:

> Air conditioning may not always be an option: with peak power demand occurring during the driest part of the year when reservoir levels are already low, hydroelectric power outages could lead to blackouts during the worst heat waves.[67]

It seems more than likely that many people will die before there is any kind of voluntary migration. During the pan-European heat wave of 2003, the death rate from heatstroke was estimated to be between 30,000 and 50,000 people.

Unfortunately, this is just a taste of what's to come. By the 2040s, even though *global* average temperatures may not yet have risen 2°C, *more than half* of the world's summers will be hotter than they were in 2003 in Europe. As a result, death tolls will rise remarkably, and our major cities will become barely habitable for three or four months of the year.[68] Concerned parents and worried seniors will leave America's cities for months at a time in the coming decades simply because of heat. America will completely lose its cool. Heat, in combination with factors like extreme water shortages and broken urban infrastructure, will encourage many city dwellers in America's largest cities not to return.

When people finally see that climate change is truly happening, truly forcing them to change their lives, they will begin to ask this question: *"Where is the safest place to go?"*

CHAPTER 6

Drought in
the Carbon Summer

*Disasters associated with climate extremes influence
population mobility and relocation.... If disasters occur more
frequently...with greater magnitude, some local areas will
become increasingly marginal as places to live.... In such
cases migration and displacement could become permanent...
[producing] new pressures in areas of relocation.*

C. B. Field, "Managing the Risks of Extreme Events" (2012)

By July 2012, the reality of climate change had been reluctantly accepted by a majority of adults in the continental United States.
Somewhat unfortunately, this acceptance was not due to careful
reconsiderations of the science involved. It followed the freakishly
extreme weather of the preceding 18 months, the hottest since record keeping began back in 1895.

At Stockholm University in 1895, Svante Arrhenius became
Professor of Physics—even though he was a chemist by training.
In the months following his controversial appointment, Arrhenius
developed an hypothesis that bridged both scientific fields with
the claim that elevated levels of carbon dioxide in the atmosphere
would act much like the glass windows of the Swedish "växthaus"
he remembered from his childhood home in Uppsala.[1]

Much later, the planetary warming effect that results from increasingly high levels of atmospheric carbon became known in

English as *the greenhouse effect*, following Arrhenius's observation. By "greenhouse" he intended to describe a mechanism that traps heat in the earth's atmosphere and thereby causes climate change. We call this change "anthropogenic," or man-made, because the increase in atmospheric carbon comes from the increase in mankind's industrial activities which take much of their energy from heat-exchange processes that release carbon dioxide during combustion. This began with the coal economy that developed after the Industrial Revolution, and it continues with the oil economy. Oil and coal are, of course, both fossil fuels.

Carbon dioxide (CO_2) is the main greenhouse gas. It remains in the atmosphere for 100 years. Concentrations of heat-trapping atmospheric CO_2 averaged 390 ppm (parts per million) in 2011. Before the Industrial Age, the average CO_2 ppm ratio (judging from ice core samples taken at the poles) was only 280 ppm, so about 350 billion metric tons of CO_2 have been added to the atmosphere since 1750. Between 1900 and 2011, CO_2 and other gas emissions increased the warming effect on the climate by 30%. According to World Meteorological Organization (WMO) Secretary-General Michel Jarraud, the CO_2 will "remain there for centuries, causing our planet to warm further and impacting on all aspects of life on earth.... Future emissions will only compound the problem."[2]

Unfortunately, the *greenhouse effect* is very badly named in both English and Swedish. Using the analogy of a greenhouse has led to the misguided belief that significant increases in CO_2 will lead to an era of much higher-yield harvests. While there may be a small increase in the growth of *some* flora, especially among undomesticated plants like poison ivy, this tiny window of improvement disappears quickly with increases in heat. Rising temperatures destroy plant life by dehydrating the living topsoil; it dies, turns to brown-grey dust, and is then carried off in wind or water. The increasingly extreme temperatures of what has been called *the carbon summer*, a prolonged geologic period of extreme global temperatures, will make the growth of any flora difficult and eventually impossible.

There is nothing "green" about the greenhouse effect.

Even more disturbing, however, is the effect of a carbon-warmed planet on the earth's methane, which is currently trapped in the frozen tundra and on the polar seabed. In seawater at extreme temperatures, methane freezes into a crystalline structure with a variety of names including *clathrate hydrate, methane ice, hydro-methane* or—much more poetically—*fire ice*. With global warming, these structures break down, releasing methane into the surrounding ocean where it boils to the surface in "methane chimneys" stretching a thousand meters from the shallowest depths of the Arctic Ocean before becoming part of our atmosphere. Methane is a much more effective greenhouse medium than carbon dioxide, and methane levels in earth's atmosphere began to spike in 2007. Significantly, the coastal temperature of eastern Siberia increased nearly 10 degrees in the first decade of the new millennium.[3]

In 2009, when I first tried to interest a publisher in this book, climate-change-denial was still in force, and less than 30% of US citizens believed the world was getting hotter and that seasonal weather was undergoing long-term transformation. Last summer, barely three years later, the number of American believers in climate change had jumped to nearly 70%. In July, Ingrid Witvoet at New Society Publishers contacted me expressing renewed interest in an old idea, then called *Brokedown Palace: Climate Change and Human Migration in North America*.

Apart from the title, what changed was the weather. Global warming is no longer a rumor about polar bears dying in the far North. It has become and remains a visible reality no matter where you live in the world. In July 2012, the *New York Times* reported, "We're now in the midst of the nation's most widespread drought in 60 years, stretching across 29 states."[4] In January 2012, 550 of the largest American cities recorded their highest monthly temperatures ever. *USA Today* reported "freakishly warm January temperatures across nearly the entire USA"; it was this heat that provided the ingredients for the 70-plus twisters that tore up towns across the Great Plains and the Southwest and killed two people in Alabama.[5]

That month—weirdly—much of Alaska had the coldest temperatures on record. And in Reno, winds of 85 mph (hurricane wind speed, according to the Enhanced Fujita scale) sparked a freak winter fire that destroyed 30 homes. In February, more uncharacteristic warmth continued in the Northeast, and, toward the end of the month, 13 people in Kentucky and Illinois were killed by more tornadoes. On Tuesday, February 28, a cold front moved eastward from the Rockies. Then on "Leap Day," these frigid mountain winds "slammed into the warm humid air [covering]...the eastern half of the nation."[6] Simultaneously, a massive blizzard brought instant winter to Colorado, dumping 52 inches of snow in one spot and covering Denver with nearly 16 inches. March brought extreme blizzards to the northwest, and even more tornadoes that killed another 39 people.

Most Americans live in the Northeast. A majority of these will remember the now-famous "summer in March," but the actual summer months of 2012 became a simmering haze of heat and drought stretching inexorably across the country like a wasting disease. This heat wave not only killed many urban residents, it also destroyed 70% of America's corn and soybean crop. The impact of the immense crop loss is still felt today in the expense of beef, pork, chicken, and milk, as well as in corn-based sweeteners—those ubiquitous ingredients found in about 80% of all supermarket products. They've played a major role in American obesity, so their increasing expense may be a small blessing.

Can We Really Blame Climate Change?

Is climate change to blame for our extreme weather and increasingly unaffordable food? The short answer is "yes, of course, sort of." But the messiness of this incomplete explanation leaves it vulnerable to misunderstanding, exposure, attack and denial. Too many of these questions have already been met with complete denial for far too long. Climate-change-denial is an even more powerful enemy than atmospheric carbon dioxide. CO_2 is a huge problem to be sure, but one that we might be able to mitigate if we all went to work on it with imagination, determination, cooperation and money. Denial, however, sits on its hands and refuses to budge.

Denial is fear. But fear is sometimes a completely appropriate response to threatening situations; denial, on the other hand, is self-destructive. It leads to death because inaction results from denial. It's a harsh truth that you can never wish unpleasantness out of existence. As the great humanist, W. C. Fields once said, sometimes you just have to, "take the bull by the tail, and face the situation." Luckily, human beings are capable of great self-control and decisiveness in the face of fear. We call this very human quality *courage*, derived from the French word for "heart." For what mankind will face in the coming years, we will all need miles and miles of heart.

The good news is that nearly 70% of us now believe the climate is changing. The bad news is that practically no preparations have been made yet. There are still many die-hard gainsayers, including 2012 presidential candidate Mitt Romney, who ridiculed the idea of climate change in his nomination acceptance speech. Mr. Romney has faded from the scene, but nonetheless we should look carefully at what the scientists say while remaining aware that their language, the language of science, is an extremely guarded one that reaches conclusions with great reluctance. Preparation is what we need. But this preparation must be precisely focused.

Speculation—the basis of any hypothesis—has no place in scientific conclusions. Scientists don't publish statements like "and now that we know that, perhaps it might be so that...." Scientists search for verifiable truths. But the widespread acceptance of established truth (which is what passes for certainty) often requires generations in order to become established (the acceptance of evolutionary theories is a good example). And in the case of climate change, we don't have the luxury of waiting for absolute certainty. Our time is running out at an alarming rate. We need to be satisfied with what scientists call the *highest probability hypothesis* (that's "best-guess," to the rest of us).

Although it would be a joy to be wrong, we simply cannot afford to bet on the wrong horse here. In the few moments left to us, we need to face the highest probability—that climate change is a time-bomb that has the potential to kill tens of millions. (Unfortunately, this is not an exaggeration; the force of Superstorm Sandy was twice the energy of the Hiroshima atomic bomb.[7]) No expense or effort should be spared in preparing for climate change. If and when a crisis arrives, further preparations will be impossible. Certain or uncertain, the only time to act is *now* since we are facing not only the impending crises of climate change but the challenges our own actions—or lack of actions—are currently producing.

But don't lose heart or your courage yet. We are a species of risk-takers whose hunches have served us well since the Stone Age began 2.6 million years ago. We may yet be able to adapt to this newest (and greatest) threat our species has ever confronted. Still, you can be sure that if we do survive, humans of 2300 will be very different creatures from *Homo sapiens*. Let's give them a name: I like the meaning of *Homo tapeinos* (the Greek word for "humble"), but it's not very catchy. Fortunately, these days, it's permissible to mix Greek and Latin roots when inventing a species' name, so let's try *Homo humilis*, the humble human being. For me, this adequately describes the adaptation we need to cultivate in order to save our children from their parents' arrogance and greed. Nature may already be performing exploratory work toward eliminating ruthless, economic competition from our species. The National Center on Birth Defects and Developmental Disabilities in Atlanta recently established that from 1979 to 2003, Down syndrome births increased 31%.[8] Two of the most common characteristics of people with Down syndrome are their loving nature and their ability to bond easily with others.

Migration Fever

In 2005, [Dallas-Fort Worth] suffered its second worst drought on record. Lake Lavon, a 21,400 acre reservoir... one of the main water supplies for Dallas, dropped to 38% of full.... That year Texas suffered $4.1 billion in crop and livestock losses. The drought continued the next year and the next. In 2008, about 48% of Texas suffered severe drought conditions and cattle producers lost about $1 billion, mostly because their grassland had turned to dust.[9]

When lives and livelihoods are threatened, people start to migrate. Once it begins in earnest, migration turns quickly into an infectious phenomenon called "chain migration" or "migration fever." Collectively, most people can be quite stubborn when they confront change. This is often a useful trait we like to call doggedness or determination. Dogged or determined people sometimes remain in place when it is difficult to live because of emotional attachments or perhaps the inability to accept failure, or to recognize mistaken beliefs. Nonetheless, very few people willingly accept the deprivation of unnecessary or enforced poverty. Commonly, people migrate when their lives or livelihoods are threatened, and also when oppression denies an exploited people fundamental material benefits (safety, freedom, prosperity, hope) that are available elsewhere. It was for these reasons that Moses was able to convince the nations of Israel to follow him into the wilderness. The King James Bible describes those migrants aptly. In Exodus 32.9, Jehovah tells Moses: "I have seen this people, and, behold, it is a stiff-necked people." And yet, despite their stubbornness, circumstances in Egypt were so desperate that the tribes of Israel followed Moses and his brother Aaron into a vast desert looking for a promised land.

During America's Great Migration, the circumstances for African American fieldworkers were similarly oppressive and exploitive, and the Promised Land was now located in the Northeast, where former field hands could find better paying jobs. From the Jim Crow-era Delta in 1910, about 300,000 field hands migrated northward when they learned about opportunities to share in the American prosperity. In the past, African American migration to the industrial north was attributed to a single, environmental factor: the arrival of the boll weevil from Mexico in 1909. But the boll weevil's role in this migration is much less significant than once thought, although it has taken a while for historians to clarify this point. The boll weevil, of course, was a cotton-devouring insect once thought to have destroyed the South's textile industry by, among other things, driving the entire black labor force out of the South. As a recent historian of the Great Migration observes, in the "1910s and 1920s [as] the [boll] weevil spread east to the older lands of the cotton belt...laborers there looked west as well as north for places to resettle."[10]

The resulting migration north numbered 300,000 African American laborers in 1910, the year following the boll weevil's arrival. The infectiousness of these numbers resulted in a chain migration which African Americans of the day called "migration fever." By the end of the war the number of African Americans outmigrating from the southern states had jumped to a million.[11] By 1930, 1.6 million southern blacks had moved to northern cities and taken up permanent residence there. As this northward movement happened, however, poorer black laborers poured into the Delta to take the north-bound migrants' abandoned jobs. So during the first phase of the Great Migration, the "number of black sharecroppers in the Mississippi Delta actually increased to its highest level ever: 77,000 tenants."[12] Until recently, this confusing demographic fact obscured the push-and-pull factors influencing the Great Migration.

The Great Migration and the migration of Mexicans to the United States since "la crisis" in the 1980s both demonstrate that a job one person considers beneath his dignity often offers the dignity of subsistence to someone who is slightly more desperate. Before the current recession, a central plank of the American Dream used to be that every successive generation of American-born offspring would unquestionably be able to do better economically and professionally than their parents—as long as they applied themselves. A similar hope is one of the most powerful pull-factors during mass migrations. It is for this reason that in coming years, young American adults—the most educated, indebted and unemployed sector of the American public—will be among the first to consider migrating to a more prosperous way of life in another country where they can start fresh and hope for improvement instead of resigning themselves to paying off the national deficit and their own student loan debts. This is especially true since the country they will look to has demonstrated a commitment to ensuring the well-being of its citizens by creating a social safety net. The desperation of these young Americans is already patent. CNN reports that

> in 1984, households headed by people age 65 and older were worth just ten times the median net worth of households headed by people 35 and younger.... Now that gap has widened to 47-to-1, marking the largest wealth gap ever recorded between the two age groups.... Older Americans are currently 47 times richer than the young.... Households headed by adults age 35 and younger had a median net worth of $3,662 in 2009...a 68% decline...compared...to 25 years earlier...[while] households headed by adults ages 65 and older...rose 42%, to a median of $170,494.[13]

Meanwhile the number of employed young adults "has dropped to 54.3%, the lowest level since...1948." Many of these people "are scraping" through America's recession "by...waitressing, bartending and [doing] odd jobs as they wait for the economy to slowly recover."[14] In *The Clash of Generations* Laurence Kotlikoff describes how America's young people alone will bear the burden of the nation's ongoing financial crisis. Kotlikoff believes the fiscal cliff that confronted the United States in December of 2012, is one side of a fiscal gap created by debt which is as wide and deep as the Grand Canyon. We have mortgaged our future and left our children's generation to pay the debts we accumulated:

> Our country's problems are real and they aren't going away on their own.... We're $211 trillion in debt, saving nothing, investing next to nothing, in most cases experiencing no real wage growth, suffering high unemployment, growing more unequal, getting older, refusing to die, balancing a

phony budget, fighting wars of pride rather than of purpose, dumping massive liabilities on our children while sustaining a "trust me" banking system poised to re-detonate. We are, in short, totally screwed.[15]

The debt now confronting young Americans is as much a push factor for future migrations as are climate changes like permanent drought. In combination with climate change, economic collapse will set people in motion. We've seen that the most human aspect of drought is that it destroys livelihoods, making migration inevitable. The Poppers showed how residents of the Great Plains began "out migrating" during the Depression and drought of the 1930s. This ongoing exodus will soon turn the entire region back to genuine, unpopulated wilderness. More recently, an intensification of the region's dryness—the recurrent droughts of the 21st century—have raised the question of causation in people's minds. The great drought of 2011 also raised the possibility that America's southern states will never return to a period of extended non-drought.

2005–2011 Texas Drought

The severity of this recent event was described by a bold climate scientist attempting to connect climate change to the 2011 drought using probability theory:

> In 2011, the state of Texas experienced an extraordinary heat wave and drought. The six-month growing season of March–August, and the three summer months of June–August were both, by wide margins, the hottest and driest in the record that goes back to 1895.[16]

Gradually it settled in people's minds that the drought that began in 2005 (and which continues today) was a harbinger of the "permanent dry" that is gathering force and speed in the second decade of the 21st century. The drought continued into 2011 without respite. As the US Dept. of Agriculture observed, the drought continued

to dominate much of the country from the Great Plains eastward, with weekly temperatures averaging more than 8 degrees [F] above normal in an area centered in Oklahoma...
[while] the Pacific Northwest and Texas received less than
2% of the normal weekly rainfall.[17]

The economic impacts of this prolonged heat wave and drought
were devastating to Texas and to the South generally, and they
provided Southerners with a powerful reason to make a realistic
reassessment of the impact of climate change on any business vulnerable to heat and dehydration. Slowly, more and more Texans
came to view the rapidly expanding population of their state with
trepidation borne of worries about sufficient water. Southerners
generally became convinced that despite occasional patches of green
during the coming years, their home states are undergoing a process
of desiccation and desertification that will forever destroy fundamental industries like cattle, citrus, cotton and even barge-freight
traffic along the Mississippi. But this collective changing of mind
was a slow process. Luís Alberto Urrea describes the mood of talk
radio shows as he drove his family home to Nebraska during the
burning summer of 2011:

> Fellow Americans spent the summer scoffing at "global warm-
> ing." It's a cycle! they said. It's snowing in _____,
> they insisted. They also seemed to be really angry that scien-
> tists acted like they were smarter than they were. But, look,
> America. It was hot. It got hotter. Then it caught fire. And
> your food died on the stalk.[18]

Not surprisingly, Americans were not convinced by their scientists;
they were convinced by their own empty wallets. As climate change
destroys a region's habitability, it simultaneously shrivels regional
economies and livelihoods. At first, determined people will remain
in places where it is difficult to live, but as the example of the Great
Migration shows, few willingly accept the deprivation of unnecessary or enforced poverty. Take that fact with the certainty that

biological existence depends on water. There is no substitute. Life depends on it and so, of course, do regional economies. Climate change will hammer this point home until even the most determined people change their minds. In recent years, there has already been a lot of preparation for collective mind-changing. Droughts in 2005 and 2008 destroyed much of the tree cover of the Southwest, making water security a vital issue in areas that had recently enjoyed enormous population growth. (Populations in Arizona, New Mexico, Nevada, Texas and southern Utah had grown freakishly fast in the decade before the droughts.) *Time Magazine* described the devastating economic impact of the 2011 drought in the final weeks of Texas's longest, hottest summer:

> Texas…is suffering the worst one-year drought on record, an average of just 6.53 in (17 cm) of rain so far this year, well off the 34 in (86 cm) it receives over a normal 12 months. At the end of July a record breaking 12% of the continental United States was in a state of "exceptional drought"…. More than 2 million acres…of farmland in Texas have been abandoned, streets are cracking as trees desperately draw the remaining moisture from the ground.[19]

No one could deny the severity of the 2011 drought, the second hottest on record for the United States.[20] That year, many Southern cities recorded their hottest temperatures, as 12% of the landmass of the United States was seared by an extreme drought. And a heat wave in July made an oven out of the entire Eastern Seaboard; many cities experienced record-breaking heat (in Newark, New Jersey, it hit 108°F). Such unbearable extremes had lasted through the entire year, so even though September was cooler, trees, brush and scrub were as dead and dehydrated as kindling; wildfires began to rage across Texas. The drought killed and then dried 500,000,000 Texan trees, basically turning them into firewood within a single year. Even when drought has parched the soil surrounding their roots, trees continue to suck or draw. What they draw into their systems, however, is not groundwater, but air, so drought-stricken trees are killed by a kind of "hydraulic failure [that] creates trapped gas emboli in the water transport system." When drought "reduces the ability of plants to supply water to leaves for photosynthetic gas exchange" the eventual outcome is "desiccation and mortality." [21]

Half a billion dead trees, of course, provide enough dry fuel for an enormous forest fire. Once the blaze began in September, the fires in Texas killed many more trees, and this was the second occurrence of wildfires since April 2011, when 1.5 million acres of Texan forests were consumed. The devastation is a frightening warning of things to come. The megafires of Texas, California and the Southwest will move northward in coming decades as temperatures ramp up. So the ugliest spawn of the permanent droughts of the 21st century aren't the dead cattle, crops or economies, they are the raging megafires that destroy homes and melt the unluckiest and most exposed human beings into ash. They leave nothing to go home to, and many fewer people to go home.

Forest Fires after Smokey the Bear

A second reason for the ferocity of these 21st-century forest fires is that such fires have been systematically suppressed for 100 or more years. A wild forest that normally supported 40 trees per acre now supports about 900 because fires have not burned away the saplings for a century. Forestry experts call this the *Smokey the Bear Effect*. As a result of this unnatural forest density, "last year (2011) 74,000 wildfires burned over 8.7 million acres [of forest] in the U.S." [22] Texas A&M Forest Service Sustainable Forestry Chief Burl Carraway reported, "wildfires…have scorched an estimated 4 million acres [of trees] in Texas since the beginning of 2011. A massive wildfire in Bastrop destroyed 1,600 homes [and] is blamed for killing 1.5 million trees." [23]

Dwight Lindsay is a retired state employee who lived in Bastrop until the wildfires destroyed his home. *The Alcalde*, the University of Texas's alumni magazine, describes the non-monetary price of extreme weather events and how their destructiveness turns people into reluctant migrants:

Dwight and his wife Marg were visiting Houston when the wildfires blew in. By the time they made it back their house was already destroyed. "It was just devastation," Lindsay says, "Everything was ash, burned rocks. We had a little cat, two years old, who had developed a real personality and I tell you this is the thing I miss most".... [Dwight said] "We had all kinds of bibles—German, Swedish, family bibles with family histories in them like people used to.... We could track our family back generations.... Margie was really into that. All of her resource materials went.... I think she misses that as much as anything." [24]

The Lindsays' painful experience demonstrates that the real costs of natural disasters cannot be accurately monetized. And the tragic destruction of their home also demonstrates that the term "drought" is misleading because although it refers to aridity or dryness and to the regionally relative absence of rain, it also implies a further danger—heat—the silent, omnipresent marriage-partner of drought that causes the excruciating phenomenon called evapotranspiration. Evapotranspiration is a four dollar word that simply describes how hot, dry atmosphere sucks the moisture out of everything. Evapotranspiration exacerbates the absence of rainfall by killing the green cover in any drought-stricken area while sucking water out of irrigation ditches, creeks, lakes, reservoirs, rivers and wells. The dryness of all droughts is exacerbated by its heat; and together, dryness and heat combine in a powerful recipe for wildfires. But since 80% of all water used in American agriculture is lost through evapotranspiration, this deficit could be significantly reduced simply by using closed piping, as they now do in Israel and in other desert farming areas.

Previously, I included Aldo Leopold's explanation of the inadequacies of scientific descriptions that simply deliver the raw facts of detached, emotionless analysis. Perhaps scientists invite people's enmity because of their "dispassionate objectivity," which some perceive as a lack of compassion. What follows is a not chilly science, but a highly poetic description of the harsh experiential realities of a Southwestern August drought from Walter Van Tilburg Clark, author of the "Ox-Bow Incident" and a Nevada native:

> The wind blew steadily and gently, but brought no rain with it. The hollow sky drank up the fogs before they could come in from the sea, and remained cloudless. At night it was filled with stars too big and too distinct.... The dry grass on the hills leaned always the same way and whispered cautiously. The river lost its strength to cut the sand-bar at its mouth, and sank away into stale pools on which the green scum spread vigorously. Boulders emerged from it into the light, and when all the water was gone, the dry mud cracked in squares and the dead fish and turtles stank. Plowed fields turned to dust, and the dust settled on the leaves of the... motionless brush along the roads. Everything alive began to suffer.... Knute declared angrily that no man could work in such weather and went off to San Francisco.[25]

In the early 1990s, British-born American poet Denise Levertov described becoming "exhausted by five years of drought" in California. The hyperthermia characteristic of drought creates profound tiredness, but it doesn't take five years to do its work. People can collapse from extreme heat after three days. Even without collapse, the effect of prolonged heat on those unused to it is debilitation. It took months for me to acclimate to perpetual daily temperatures of 110 or more in Saudi Arabia. Not surprisingly, the Saudi government was extremely enlightened on this issue. The king subsidized air conditioning and considered it a basic human right. Even those who lived in the desert in tents were universally entitled to free air conditioners and to electrical power hook-ups that ran them. I found this out on the first weekend of my stay when I got lost during a Hash House Harriers desert run. With characteristic hospitality, I was rescued by Bedouin who took me back to a cluster of tents fed by thick green and white (colors of the Saudi flag) striped electrical cables. Eventually, they drove me cross-country back to my hotel in Al Khobar in an armada of Nissan Patrols. But before this happened, one of the men's fathers took me into his cool, dark tent and offered me a frosted bottle of Italian limonata. "Drink these during haboob," he insisted. "Most refreshing."

In Chapter 1, I described my encounter with the dust storms of Saudi Arabia, but the ongoing drought that began in 2005 brought similar storms to North America. By July 6, 2011 the grey storms were as big as the biggest dusters of the Dust Bowl and were regularly turning day to night in cities like Phoenix with 69 mph winds which blew in from the horizon looking like a brown tidal wave 100 miles across.

Meteorologist Jim Andrews of Accuweather.com believes that the same downdrafts that thunderstorms on the Great Plains turn into tornadoes also "kick up dry, loose sand on the desert floor creating a wall of dust that travels outward spanning a much larger area than the thunderstorm itself." [26] Americans used to call these storms "dusters," but as a sign of increasing globalization, most news outlets now call them by their Gulf Arabic name *haboob*. In English, haboob translates roughly as "the blowing," or perhaps "the/a big blow." In any case, haboobs and tornadoes are intensifying in the United States. In Blackwell, Oklahoma, on October 21, 2012, a haboob with a storm front two miles across closed the town and stopped all traffic on I-35 while causing about a half million dollars in damage. [27] The next day, reporter Beverly Bryant generously took a long time over the phone to explain to me the nature of the dust storm's damage:

> People had just turned over their dry, empty fields.... It's a bad, old-fashioned practice. Twenty miles of ditches around the county are now filled with about four feet of dried-out farmer's topsoil. It's going to take weeks to dig them out." [28]

In the United States, during dusters and droughts, the use of air conditioners soars: Power plants struggle "to meet record demand for electricity in the face of 100-plus-degree...temperatures day after day." [29] Few people realize that all electrical production, not just hydro-power, depends on water. For nuclear stations, a constant supply of freshwater is needed to reduce the heat of the fuel rods and prevent meltdowns. So generating any kind of electricity always produces enormous heat; the American electrical grid withdraws much more water than agriculture does. During droughts, "power plants [compete] for water that farmers want for their devastated crops."[30] Significantly, the *New York Times* reported: "During the 2008 drought in the Southeast, power plants were within days or weeks of shutting down because of limited water supplies."[31] World Meteorological Organization (WMO) Secretary-General Michel Jarraud put the case strongly, arguing that the problem of drought is now a permanent one:

> Climate change is projected to increase the frequency, intensity, and duration of droughts, with impacts on many sectors, in particular food, water, health and energy.... We need to move away from a piecemeal crisis-driven approach and develop integrated risk-based national drought policies.[32]

The deadly combination of drought and heat that turned Texas and half a dozen other states into saunas in 2011 and 2012 also primed them for wildfires. As of mid-2013, many states have still not recovered and, over the long-term, as Michel Jarraud has said, there will be no real recovery unless global preparedness becomes a reality. Like California, Texas will soon become a *push-out* rather than a *pull-in* state. And in fact, the "permanent dry" will affect the livelihoods of everyone in both the tropics and the sub-tropics. Regionally, the fishermen of Galveston suffer because low water levels have turned Galveston Bay so saline that it will not support fish; Southern homeowners suffer because the value has gone out of much of their investment, as concrete foundations of houses in areas of profound drought crack in the bone-dry ground; farmers suffer because 94% of Texas pasture and rangeland is now rated as "poor or very poor." Wheat farmers in Kansas have seen their crops fall by 25%; in Texas and Oklahoma, they've fallen to fully *half* their normal levels; cotton farmers suffer the same way. The 2011 drought cost more than $4 billion in direct losses, and the eventual economic loss could easily be twice that amount.[33] 2012 was no better. "Blistering heat" that year "destroyed 45% of the corn and 35% of the soya...crop in the worst harvest since 1988."[34] These figures are terrible, of course, but the worst news is not that *it ain't over yet*.[35] The worst news is that it ain't *ever* gonna be over. In coming decades, Texas will get as hot as a toaster—and it will stay that way.

In addition to outright heat battering the economy, there are also some *very* unexpected economic repercussions. The deep channel of the Mississippi River, for example, has been used for decades to lower the cost of shipping raw materials and manufactured goods from north to south. In general, barge fuel efficiency is greater than that of rail. Once underway, a barge (especially one headed downstream) requires very little fuel for navigation. It takes 60 trailer trucks to carry the cargo of a single barge.[36]

There is a lot of freight on the Mississippi: "about $180 billion worth of goods move up and down the river [annually] on barges, 500 million tons of...basic ingredients for...the US economy."[37] For example, in the short corridor of the river between St. Louis and Cairo, the wide variety of materials shipped on the river amounts to about eight million tons of grain, coal, steel, petroleum and other goods.[38] Mississippi barges move nearly 60% of US grain exports, 22% of its petroleum and 20% of its coal.[39] But in August 2012, the ongoing drought shrank the mighty river to a narrow, shallow channel. In many places where the Mississippi's width was ordinarily three miles, it shriveled to three-tenths of a mile. In the most extreme spots, it was 13 feet below normal depths. At Vicksburg, where the river once proved such a stubborn obstacle for General Grant, the Mississippi was 20 feet below average. Barge traffic became excruciating and slow. The delays cost an average of $300 million per day. A significant portion of this money was spent on the tugboats required to pull the mired barges free. The vessels cost $10,000 a day to operate.[40]

By November 2012, the problem had become much more severe as lack of rain squeezed the channel, narrowing it to just a few hundred feet in some places. River depth bottomed out also, declining to 15 to 20 feet lower than normal, making the Mississippi only 13 feet deep in many places. The minimum depth for barge traffic is about 9 feet because submerged rock pinnacles make it impossible for barges to pass.[41] Historically, an "ice bite" makes December and January the months when water level is lowest because tributaries freeze after a dry fall.[42] In a drought year, of course, this effect is much more pronounced, and in the winter of 2012, the level of the river fell to historic lows, forcing shippers to find alternate methods for moving their goods.

When the steady decline of Mississippi shipping was factored into the cost of southern agricultural products, a bitter truth became obvious by the late summer of 2011: "From beef prices to the cost of a pair of socks, the...drought of 2011 [left] a mark on Americans."[43] 2012 only increased the economic damage of this crippled shipping channel. For residents of America's drought-stricken South, the devastation of recurrent drought became a convincing proof of climate change late in 2011.

Ironically, the drought was also the phenomenon which finally enabled scientists to connect the extreme weather events happening in North America to climate change. Today a majority of Southerners share the belief of the scientific community that the deserts of northern Mexico are creeping into the southwestern United States as the tropics themselves move poleward under the continuous influence of global warming.

It happened this way.

SMOKING GUN.

A lot of people are asking about the connection between global warming and the arid landscape in the Midwest. Is climate change causing this drought? Didn't the United States suffer worse droughts in the past? And what will happen if the planet keeps heating up? The short [answer] is this: Droughts have multiple causes. The United States has suffered worse droughts in the past. It's not yet clear whether we've reached the point where global warming is making droughts worse again, at least in North America. But evidence suggests that droughts will become more intense in many parts of the world if the planet keeps heating up—a trend that could disrupt the world's food supply.[44]

One scientific report of 2012 reads, "The attribution of single extreme events to anthropogenic influences is *challenging*." Another 2012 report similarly reads "The attribution of single extreme events to anthropogenic climate change *remains challenging*." Among scientists, identifying the smoking gun of climate change has proved a very tricky problem. To break it down, their first problem is to describe a meteorological phenomenon accurately: single weather events—tornadoes and hail storms, for example—can resist these descriptions because they appear and disappear very quickly. These single events are also as difficult to grasp as tapioca pearls. Their relative intensity, and seasonal occurrences make them very slippery events that continuously elude your grasp. Once an accurate description is developed, however, the second problem is how to account for the phenomenon *theoretically*—within the confines of what is already known to be true. The third step is then to prove or disprove a new, descriptive theory. This step-wise approach breaks the natural world down into bite-sized chunks of controllable phenomenon-based data: descriptions, theories, proofs. So a lot of effort goes into nailing down the cause of a single extreme weather event.

Unfortunately, many natural phenomena have more than one cause. For this reason—until very recently—scientists claimed that "it simply was not possible to make an attribution statement about an individual weather or climate event."[45]

The obstacle of attributing single extreme weather events to climate change was only overcome once scientists lightened up on their fixation with absolute, certifiable truth and began to settle for more descriptive probabilities. As one group of scientists write, it is now widely accepted that attribution statements about individual weather or climate events are possible provided proper account is taken of the probabilistic nature of the attribution. One analogy of the effects of climate change on extreme weather is with a baseball player…who starts taking steroids and afterwards hits on average 20% more home runs in a season than he did before…. You would be able to [say]… that all other things being equal, steroid use had increased the probability of [home run hits] by 20%.[46]

Another, more topical analogy might be the comparison of recent extreme weather events to Lance Armstrong's cycling victories. Armstrong was, by all accounts, a competent, but not extraordinarily gifted international-class cyclist until he began blood-doping. Once he began using his "blood cocktail," he began winning races and was able to cycle from victory to victory as coolly and slyly as James Coburn cycled out of France in *The Great Escape*. Eventually, Armstrong won seven Tours, going home with two more gold medals than any of the three previous, non-doping champions who in their entire careers could only manage five wins.[47]

When viewed in this context, Armstrong's sixth victory exceeds probability by nearly 30%, while the likelihood of a seventh victory is incalculable and completely off-the-charts—indicating something extraordinary was at work: either a favorable mutation in Armstrong's blood cells (as his publicists claimed) or (as we now know) the ongoing chemical alteration of his metabolism. In future years, the analogy between Armstrong's chemical fraud and climate change may become gruesome as it extends to the impact of prolonged chemical-abuse both on Armstrong's body and on our global metabolism, the climate on which we all depend.

The application of probability theory to meteorological phenomenon simplifies things enormously. The statistical principle of "stationarity" comes into play in establishing such probabilistic attributions. In the past, the frequency of extreme events has been roughly predictable because their occurrence has maintained more or less the same probability over time. But climate change skews such predictions by increasing the frequency, intensity and probability of such events. A dry spell might be called a once-in-a-century drought because, as severe as it was, it only occurred once every century. The last such drought was in 1950–1956. The current one began in 2005 and is still going strong in 2013, so this time there were only 49 years between once-in-a-century drought events. And the frequency and intensity of other events have also increased. They are no longer stationary and predictable processes. New York Governor Cuomo put the matter succinctly describing the disastrous impact of Sandy on the TriState region in October 2012:

> There's been a series of extreme weather incidents. That's not a political statement. That's a factual statement. I said to the president kiddingly the other day, [that] we have a 100-year flood every two years now.[48]

Theorists of climate change predicted that such things would happen, but the devil is in the details. Even the strongest statements by teams of leading climate scientists (two examples appear below) offer little specificity or help for the ordinary reader trying to stay informed:

> (1) It is very likely that there has been an overall decrease in the number of cold days and nights, and an overall increase in the number of warm days and nights at the global scale.
>
> ...
>
> (2) It is virtually certain that increases in the frequency and magnitude of warm daily temperature extremes and decreases in cold extremes will occur in the 21st century at the global scale. It is very likely that the length, frequency and/or intensity of warm spells or heat waves will increase over most land areas.[49]

Using a much more localized probability model, however, the editors of the *Bulletin of the American Meteorological Society* were able, in July 2012, to connect the dots of several extreme weather events directly to global climate change. One of these was the Texas drought of 2011. They found that

> extreme heat events were roughly 20 times more likely [than they had been 40 to 50 years ago].... This suggests that conditions leading to droughts such as the one that occurred in Texas in 2011 are, at least in the case of temperature, distinctly more probable than they once were.[50]

So, even though it is smudged and faint, nonetheless, "the fingerprint of human activity has been detected in the recently observed warming on global and continental scales."[51] The most important extreme event that contributed to this change was the 2003 heat wave in Europe that caused somewhere between 30,000 and 50,000 excess deaths.[52]

For governments, scientists, citizens, healthcare professionals, and emergency service workers, the solid attribution of extreme weather to climate change became a crucial issue following this heat wave. In France alone, 14,802 people died—a figure that represents "more than 19 times the death toll from the SARS (Severe Acute Respiratory Syndrome) epidemic worldwide."[53] Governments needed a hard, factual answer to the question of whether they could expect recurrent similar disasters. And if so, how often might they be expected to occur?

The statistical perspective of extreme weather events already illuminated the risks insurance companies faced through the increasing frequency of extreme weather events. For some, the view that climate change is an omnipresent and intensifying threat is a cold, hard, mathematical fact as dispassionate and unforgiving as an actuarial table. Since 1980, Munich Re and other insurers have noted that estimates of annual losses from weather and climate-related disasters have skyrocketed from a few billion to more than $200 billion (in 2010 US dollars). The highest losses so far were incurred the year of Hurricane Katrina, but Superstorm Sandy (2012) runs a very close second, and its total economic impact includes shipping disruptions that continue into 2013. But we should not forget the losses suffered by the Lindsays, whose home was burned to the ground by wildfires in Bastrop, Texas. Our lives are not made rich by the value of our material and replaceable possessions, but by the deeply personal significance of people and things in our environment. For this reason,

> loss estimates are [always] lower bound [a technical term meaning conservative or overly cautious] estimates because many impacts, such as loss of human lives, cultural heritage, and ecosystem services are difficult to value and monetize, and thus they are poorly reflected in estimates.[54]

In Texas, the drought of 2005–2013+ has crippled the economy. And researchers have found that such events are now 20 times more likely than they were in the 1960s. Moreover, there is little prospect for relief, mitigation or respite. A variety of studies predict gradually worsening extreme heat over the Southwest in coming years. The "mean" climatological state of Texas and the Southwest during future years will always be "moderate drought conditions," and these will extend over most of the western United States. But in the La Niña years (which will increase in frequency and intensity in decades to come) drought conditions will become even more severe than they have been during the years 2005–2013.[55] Another study by David Easterling found that the United States has been getting hotter since 1950 and that this gradual desiccation had dried out soils, preparing the Southwest region for desertification.[56] Columbia University Professor Richard Seager described the outlook for the region as a "permanent dry."[57]

There is only one solution when semi-aridity becomes desertification and the human habitability of a region is radically reduced. "March or die," said the horrible Foreign Legion sergeant played by Telly Savalas in Beau Geste. This is the same strategy the Anasazi adopted centuries ago. It is the same strategy observed by the Poppers on the Great Plains. In 2012, the IPCC observed:

Disasters associated with climate extremes influence population mobility and relocation, affecting host and origin communities.... If disasters occur more frequently and/or with greater magnitude, some local areas will become increasingly marginal as places to live or in which to maintain livelihoods. In such cases, migration and displacement could become permanent and could introduce new pressures in areas of relocation.[58]

Unfortunately, down South, none but the already thirsty native Texans have awoken to the fact that there is not enough water to support the ever-growing population in the region. There is a lag time as the prosperity that freed most Texans from state taxes fades in the harsh reality of concerns about heat and water. The first iron-clad rule of a warming world is "there is no substitute for water." For this reason, I believe that very soon the people of the Southwest will begin looking for cooler climes and higher ground because anthropogenic climate change will only continue—in the words of my west-Texas cowboy friend, Jack Everton—"wuppin' ass 'n takin' names."

A migration in the coming decades would move residents of the South into the opposite corner of the United States from the chosen destination of the Great Migration a century ago. North of Oregon and Washington, the combined area of Alaska and British Columbia beckons, with about 1,000,000 square miles and vast resources of water renewed constantly by rain from the Pacific's westerly winds. Population density in Texas has nearly reached 100 people per square mile, but in most of British Columbia (outside Vancouver) and in Alaska, the density figures are single digits. It will take several decades, but the migration I foresee will be definitive. The latitude and rains of America's Northwest Coast, British Columbia, and Alaska will present a last hope to desperate people in desperate times.

They will come. They will have to.

CHAPTER 7

Wind and Water

You can't have a storm offhand,
like somebody took a notion and
decided a storm would be right
handy to come off now and here.
The moan and lash of the wind came out of a place
nice for them, nice for their growing.
The anger of the waters lay breeding, spawning,
pent up and ready to go.

CARL SANDBURG, EXCERPT FROM
"STORMS BEGIN FAR BACK" (1946)

These days, the storms of mid-winter or high summer have acquired a new ferocity. As seasonal boundaries shift, these tempests no longer obey the old parameters of occurrence. Blizzards come any time, and they now have names like "Superstorm Nemo," which visited the northeastern United States in February 2013. Tornadoes too, now regularly come out of season, arriving even in the depths of winter and ranging throughout the Great Plains and the Southwest. Both the timing and location of hurricanes have changed. Now, they often arrive in late fall and have moved steadily northward. Simultaneously, the ranges of American songbirds have moved northward, as have zones of agricultural production. In the second decade of our new century, every year has been a warm year, with spring's arrival occurring 10 days earlier in the northern hemisphere. But 2012 was an exceptional year, and spring arrived 20 days (three weeks) early. As New York Governor Andrew Cuomo observed, "Anyone who thinks there is not a dramatic change in the weather patterns is denying reality."[1]

The most obvious dramatic proof is the change that has taken place in tornadoes. Tornadoes, which are mainly an American phenomenon, are a benchmark for climate change. A putative tornado appeared very far north in the Canadian province of Ontario in April 2013. In the United States, seven days of tornadoes (30 twisters of varying intensity) were confirmed in October 2012. Then there were four more tornado days (involving seven more storms) in November. The next month saw eight more days scarred by these whirling seasonal anomalies. Altogether, there were 46 tornadoes in December, including the 27 storms that spoiled Christmas Day for residents in towns and cities across Alabama, Louisiana, Mississippi and Texas.

Old school meteorology holds that the magnitude of a tornado is best measured by the maximum wind speed sustained for a single minute. A minute used to be a long time for a tornado—most appear and disappear in just two to four minutes. Larger and more forceful tornadoes last between 20 minutes and one hour, but these used to be rare events. To date, there have only ever been eight tornadoes with the highest rating (EF5). But—and get this—*one half* of those EF5 killer-tornadoes occurred the late winter and spring of 2011.

Simply put, tornadoes are getting much worse.

It is "wind intensity" that scientists use to classify the force of tornadoes using the "Enhanced Fujita" scale. Unfortunately, this scale does not describe storm damage very well, even after "enhancement." The major problem is that a small tornado can cause a crazy amount of damage in a densely populated area *if it lasts a long time*. On Christmas Day 2012, for example, the intensity of the storms varied between EF0 and EF3, or between "hardly significant" and "moderately disastrous," but the best-known storm of the day was rated merely as a "persistent EF2 twister"; that tornado turned out the Christmas lights in downtown Mobile as it crisscrossed the city in what residents described as "waves." The whole process lasted about two hours. During that time, the tornado felled whole streets of mature live oaks, leveled a school, demolished several homes, and deconstructed a Protestant church which had been overflowing with worshippers only a few hours before.

Characteristically, before the Mobile storm hit each neighborhood, residents felt their ears "pop" as the barometric pressure suddenly dropped. This was their only warning; the telltale "freight train noise" that indicates a storm is upon you, leaves little time to get to real cover. You need to find a makeshift foxhole wherever you are.[2] In the southeastern United States, if you hear a freight train behind you where there are no tracks, you should start to move (and pray) quickly, just as Essie Hendricks, the manager of Peebles Department Store did in Yazoo City, Mississippi, on the morning of August 24, 2010. With very little warning, a tornado ripped away the building around her:

> I saw it.... It was a cloud of black smoke coming up the hill. And I could hear it. It sounded like a thousand trains. In half a second I had enough time to get under the front desk. And by that time it hit. It sounded like the building was going to explode...you just heard...rumbling! Banging!...hitting! It got so dark. Tables flew.... Man, you could hear the glass breaking. All the cologne bottles shattered.[3]

Essie's desk proved a good bet at the last second. If you have a bit more time, you should cover yourself with a mattress (good) or a thick blanket (chancy) in order to protect yourself from the hailstorm of windblown debris that always accompanies a tornado. Unfortunately, there is always very little time, and if you're especially unlucky, you will simply become a witness to the tornado's power, as Will Keller famously did when a twister overtook him in his Kansas pasture in 1929:

> There was a circular opening in the center of the funnel...
> 50 to 100 feet in diameter...extending straight upward for a

distance of at least one half mile, as best I could judge under the circumstances. The walls of this opening were...rotating clouds and the whole was made brilliantly visible by constant flashes of lightning which zigzagged from side to side.... I noticed that the direction of rotation of the great whirl was anticlockwise, but the small twisters rotated both ways—some one way and some another.... The opening was completely hollow.[4]

Keller's nose-to-nose account is remarkable. Very few survive such close encounters. Nonetheless, 50 years before Keller's experience, John Park Finley, a pioneering American meteorologist, studied and chased a lot of tornadoes across the Great Plains. Lieutenant Finley was a member of the Army Signal Corps, which was troubled by the impact tornadoes had on America's telegraph network. Telegraph poles and wires were a major communication expense for bi-coastal trade in late 19th century. None could withstand even the mildest tornado. In 1887, Finley wrote that, judging from his own experience, tornadoes can

never be effaced from the memory of the actual observer. There is an awful terror in the majesty of the power here represented, and in the unnatural movement of the clouds which affects animals as well as human beings.... [Their] terrible violence...makes the earth tremble...and men's hearts quake with fear.[5]

Today, many Americans accept intense, frequent, longer-lasting and unseasonable tornadoes as proof of a changing climate, although they explain such climate change in radically different ways. By the fall of 2012, well before Superstorm Sandy, the number of "Americans [who] believe[d] that the severity of recent natural disasters is evidence that we are in what the Bible calls the *end times*" had fallen to a little "more than one third (36%)."[6] The declining numbers of these evangelical end-time believers correspond to what some participants call "a collapse in American evangelism" that is signified by the lack of evangelical impact on the presidential primaries, the legitimization of same-sex marriage in several states, and a state legislature's rejection of an amendment to restrict abortion.

In 2011, American evangelical leaders universally reported "a greater loss of influence than church leaders in any other country."[7] Climate change has proven to be an important element in the decline of evangelical influence because church leaders lost swaths of credibility *and* followers during the powerful and destructive meteorological events of 2011. As European insurance giant Munich Re would soon report, weather-related (insurance) loss events "quintupled over the last three decades.... Nowhere in the world is the rising number of catastrophes more evident than in North America."[8] In 2011, the expense and devastation of climate change made itself felt in America, and the first inkling of this change came with the tornadoes.

The 2008 Super Tuesday Tornadoes

Ibom (Tornado-Giant) came like a round black cloud
with his feet on the ground and his head in the sky.
He came dancing around and around...at every step
he would smash down trees and kick them from his way.

JAMES R. WALKER, MD. *LAKOTA MYTH* (1983)

Back in 2007, the UN's International Panel on Climate Change (IPCC) reported cautiously that there was "insufficient evidence to determine whether [new] trends exist in small-scale phenomenons such as tornadoes, hail, lightning and dust storms." A few months later, in January 2008, a cluster of more than 80 twisters, now famous as the "Super Tuesday Tornado Outbreak," caused approximately $850 million in insured losses—and many fatalities.[9] During the presidential election debates, these Super Tuesday storms led to a fiercely partisan media imbroglio about the validity of climate change. But four years later, on Dec 25, 2012, blizzards created instant *White Christmas* throughout the northeastern states. On the very same day, tornadoes raged across a superheated Southeast.

This time, a majority of Americans assumed the cause of both events was global warming.

It was not the persistent advocacy of scientists that convinced the majority of Americans climate change was happening. The about-face in public opinion started as public debate about the causes of the winter tornadoes that occurred on February 5 and 6, 2008. In 2008, *winter* tornadoes were generally regarded as rare, once-in-a-generation-events. Over the next four years, the severity and erratic timing of these brief, destructive and very poorly understood meteorological disturbances would cumulatively contribute to the American's acknowledgement that climate change *is* quite real. This change was possible mainly because tornadoes (together with thunder and hail storms) create 57% of all catastrophic (insurance) losses—losses which have increased remarkably in recent decades.[10] Once it was clear that insurers' pocketbooks were increasingly being affected by these events, the people and governments paying insurance premiums began to acknowledge that climate change was in progress. So by the summer of 2012, the *bumfuzzle*[11] of climate-change-denial finally began to lift. Most Americans (almost 70%) were convinced that rapid meteorological change was really happening *and* that mankind had likely caused it. The proof of anthropogenic climate change was right in front of us in the frequency and intensity of strange and extreme weather events. As things like tornadoes grew stronger and visited us earlier each year, North America's acceptance of the reality and dangers of global warming took hold. This change in our belief system is a necessary first step before Canadians and Americans begin to prepare for an impending meteorological holocaust. (Widespread acceptance of migration as an appropriate defensive strategy will be one of the last stages of this preparation. Although it may take a decade or more to get there.)

Fans of James Lovelock or proponents of the Gaia principle might find confirmation of their belief in a Global Spirit in the fact that meteorological events have forced their way into so many important recent US elections. In 2008, as 24 states held their primaries, 87 tornadoes appeared across the Great Plains and caused about $1 billion in damage during the 15 hours that stretched from before sunset to dawn of the following morning. Nighttime tornadoes are considered the most dangerous because most people are asleep and don't get media warnings or sense the tornado's approach. On Super Tuesday, the worst of these storms hit big towns in the red states, including Memphis, Jackson and Nashville. There were 57 deaths. Since 1985, no cluster of tornadoes has claimed so many lives. In fact, the worst American tornadoes had all happened before 1954; for a long time, this provided useful ammunition to the army of climate-change-deniers funded by big oil. These paid hacks blogged and published a steady stream of discouragements undermining the belief of American voters in one of the major opinions of liberal scientism. But it was impossible to ignore or belittle this unusual cluster of killer storms, as veteran Tornado Chaser Tim Samaras writes:

> In [the] southern states on Feb. 5 and 6, 2008...[a]...savage, long lasting storm complex spawned...five rare and ruinous EF4s which killed 57 people.... Together, the storms caused more than $500 million damage, making the tornado outbreak the fourth costliest in US history.... It was very nearly the dead of winter...violent tornadoes are usually creatures of the warm, wet turbulent atmosphere of spring or early summer.... Only one other tornado outbreak in the past century had killed so many people so early in the season...an Arkansas tornado...in 1949.... Another strange thing was that although tornadoes usually don't make contact with the ground for more than 10 or 15 minutes, many of these twisters were deadly "long track" tornadoes that buzz-sawed across the surface of the earth for enormous distances. One in Arkansas left a damage track 122 miles long, an all-time record.... One more oddity was that the whole South had been basking in record-breaking summerlike warmth at the time. It was 75° in Little Rock...Arkansas that day.[12]

In coming years, this combination of extraordinary winter temperatures and violent tornadoes became a disturbing and recurrent seasonal news item. Exactly one year after the Super Tuesday storms, one headline read "Rare Tornado in Winter Kills Nine in Oklahoma."[13] But the repetitive pattern was not yet recognized when John Kerry told a TV interviewer that "the weather service [says] we are going to have more and more intense storms…[and] insurance companies are beginning to look at this issue. [They] understand this is related…to the warming of the earth."[14]

Kerry's pronouncement let climate change out of the electoral bag, but the publicity surrounding this liberation did very little for his candidacy. He was roundly mocked as a fool or a liar by right-wing bloggers who claimed that any attempt to connect winter tornadoes to climate change was a democratic "hoax." In the intensity of the ensuing battle, the objectivity of both sides was compromised.[15] The upshot in 2008 was that no one knew if there was a real connection between climate change and the new phenomenon of intense, frequent and longer-lasting winter tornadoes. Afterward, climate change remained a toxic political football. Avoiding it required the best footwork politicians possessed. This unwillingness to confront the problem of climate change contributed to a widespread lack of clarity about the issue. Things were not helped by the responsible scientific position of the day, which refused to commit itself about a connection between extreme storms and climate change. As Chris Mooney wrote in *New Scientist* in August 2008: "There is still not enough data to make a firm" conclusion about the impact of climate change on tornadoes.[16]

A tentative way of connecting climate change to the new-fangled tornadoes of the 21st century would eventually be suggested by two Stanford meteorologists. The February 2011 issue of *Geophysical Letters* features a 30-year analysis by Christine Archer and Ken Caldeira, who conclude that climate change is forcing the jet stream higher and closer to the pole in both the Northern and Southern Hemispheres. They found that, as the northern jet stream moves poleward by approximately 125 miles per decade, it simultaneously increases in elevation by 5–25 meters; during that same period, jet-stream wind speeds also increased by about 1 mile per hour.[17] Their analysis confirmed the findings of two other studies published well before the Super Tuesday Tornado Outbreak.[18] All of this work demonstrated conclusively that the American jet stream is moving north. This phenomenon is significant because it confirms precisely the predictions of global warming theory. Without specifically mentioning tornadoes, Archer and Caldeira conclude:

> These changes in jet stream latitude, altitude and strength have likely affected and perhaps will continue to affect the formation and evolution of storms in the mid-latitudes and of hurricanes in the sub-tropical regions.[19]

Archer and Caldeira's findings are confirmed both anecdotally and geographically. The late tornado chaser Tim Samaras had a persistent intuition that the majority of tornadoes are now moving northward. Although not a scientist, he felt compelled to put this into print:

> I've been storm chasing for around 20 years, and it does seem to me that Tornado Alley is moving a little bit north.... We seem to be finding fewer tornadoes down in Oklahoma and Texas, and more up in Kansas, Nebraska, and South Dakota.... That's just anecdotal evidence, not real scientific evidence.[20]

The possibility that the northward creep of the jet stream impacts the creation of tornadoes should not surprise anyone. American tornado-formation depends on a unique geography that makes it the world capital for tornadoes. Cheryl Crow, who grew up in this region, describes it as "Tornado Central."[21] More often it's called *Tornado Alley*. Although tornadoes form in other countries, they occur with considerably less frequency than in the United States. On the Great Plains, tornadoes are a common occurrence because the Rocky Mountains, the Gulf of Mexico, and the jet stream combine in a geophysical cauldron of storm activity right where the

Plains Indians once venerated "Wakinyan," or Thunderbird. On the Central Plains of the United States, the supercell thunderstorms that produce the most and the longest-lasting tornadoes occur with greater frequency than anywhere else in the world. The ingredients for a supercell are strong vertical wind shear, high values of buoyancy, moisture, and heated ground. The key is heat:

> If air near the ground can be easily heated enough or lifted just enough to trigger a storm, it is likely that storms will form almost everywhere. Then destructive interference between neighboring cells is very likely.[22]

In his wonderful book, *Tornado Alley*, Howard Bluestein explains the complex relationship between the geophysical forces that give rise to supercell thunderstorms that, in turn, bring on tornadoes. Basically, three massive sources of wind (the Rocky Mountains, the Gulf of Mexico, and the jet stream) combine *during periods of extreme heat* to create the vertical wind shear that makes a supercell storm (the Thunderbird). Such storms *sometimes* lift the pounding, horizontal surf-like rotation of a tightly wrapped, whirling precursor wind into its vertical life as a relatively small tornado. But on other occasions, hot moist updrafts intensify inside the supercell storm itself. These accelerate if a colder force—called the *rear flank downdraft*—penetrates the updraft as it descends. Together, these supercell air currents combine to produce a violent swirling internal wind within the supercell. Sometimes, the resulting "mesocyclone" then becomes a powerful, long-lasting tornado.[23]

The genesis of tornadoes is a complicated and poorly understood process. But the formation of supercell storms is a fundamental requirement to their creation. Bluestein describes the interaction of geography and meteorology that combines to create supercell storms on the Great Plains:

> The jet stream...strong high-level winds...west to east... prevails during the spring. Since winds at the surface are usually much weaker...the lower half of the troposphere is likely to have strong vertical wind shear. As air flows east-

ward across the…mountains, it sinks, is compressed, and warms…. The lee of the Rockies is associated with relatively low surface pressure. With this low pressure just east of the mountains…a pressure gradient force causes air to move westward…. After several hours, the effect of the earth's rotation is felt. Eventually a state of [geostrophic] balance is almost achieved between the Coriolis force and the pressure-gradient force. The air then flows northward, bringing moisture and heat from the Gulf of Mexico to the central Plains. Disturbances…associated with relatively cool air aloft, propagate along the jet steam from west to east. With relatively cool air aloft and warm, moist air below, the environment is potentially unstable—if low-level air can become saturated and buoyant. The disturbances are associated with regions of rising motion. The rising motion cools off the air aloft and may deepen the layer of moist air at low levels. This results in a preconditioning of the air along the air mass boundaries…where even stronger upward motion can trigger storms…. Enhanced vertical shear accompanies the disturbances.[24]

Of course, the more supercell thunderstorms that occur in a single year, the more likely tornadoes (of either kind) become. Based on no better evidence than his eyesight and good sense, the previously mentioned Lieut. Finley was able to deduce—126 years ago—that heat was every tornado's active ingredient: "The time of day, the time of year, and the peculiar hot and stifling condition, indicate that heat is the physical agent developing the tornado."[25] In La Niña years, extreme heat is especially common, so with global warming, the phenomenon has intensified. La Niñas now come more frequently and last longer than El Niños. And now, such heat comes much earlier in the year, just as it did on the Great Plains in January and February of 2008 (although it wasn't recognized at the time). A February 2013 report from the US Global Change Research Program concludes definitively that annual temperatures in the United States are hotter than ever before:

Frost-free season [growing season] length has been gradually increasing since the 1980s.... During 1991–2011, the average frost-free season was about 10 days longer than during 1901–1960. These observed climate changes have been mirrored by changes in the biosphere, including increases in forest productivity…length of the growing season, and length of the ragweed pollen season. A longer growing season can mean greater evaporation and loss of moisture through plant transpiration associated with higher temperatures so that even with a longer frost-free season, crops could be negatively affected by drying. Likewise, increases in forest productivity can be offset by drying, leading to an earlier and longer fire season and more intense fires.... In a future in which heat-trapping gas emissions continue to grow, increases of a month or more in the lengths of the frost-free and growing seasons are projected across most of the United States by the end of the century.... These increases are projected to be much greater than the normal year-to-year variability experienced today.[26]

In 2012, a study attempting to explain the statistical probability of recent extreme weather events concluded that "extreme heat levels were roughly 20 times more likely in 2008 than in other La Niña years in the 1960s."[27]

Heat is crucial to the creation of the supercell storm whose internal winds create the most deadly form of tornado. But heat is also responsible for the vertical wind shear that can *loft* one end of the tumbling horizontal precursor of the other, most common kind of tornado, allowing it to connect with the storm cloud above. This is a highly chancy, two-step and nearly miraculous meteorological development; it's as though a powerful updraft suddenly stood an entire ocean-wave on its end. It depends on an abrupt boundary between the wind shear and its surroundings, and on the development of a tornado-precursor with sufficient length and power to withstand the upward movement that the wind shear causes.

Recognizing the dependency of both kinds of tornadoes (large and small) on the supercell storm above it, the Lakota Sioux identified all tornadoes as the troublesome child of an illicit union between Inyan (Rock, the first being) and Unk (Passion). The Sioux believed that Thunderbird, Inyan's rightful partner, seized Tornado whenever he broke out of his banishment, and dragged him back to the ocean realm where his mother lived in exile. They called the one-eyed tornado *Ibom* (Giant).[28]

In the years since the Super Tuesday Outbreak, tornadoes have worsened considerably. The 57 fatalities of those 15 hours in 2008 pale in comparison to the more than 300 killed by the cluster of storms that occurred April 25–28, 2011. This tornado outbreak included at least three rare monster tornadoes rated at EF5. These late-April storms began eight days after another outbreak killed 26 people in Virginia and North Carolina. The devastation to property was considerable in small towns like Hackleburg, Alabama, where whole tracts of housing were reduced to matchsticks. Tellingly, all of the April 2011 tornadoes were accompanied by a heat wave that spread across the South. On April 26 in Laredo, Texas, temperatures in excess of 110°F accompanied violent winds, spreading wildfire and death across 1.5 million acres (destroying, among many other things, the Lindsay home discussed in the previous chapter). A month later, on May 22, another EF5 tornado killed 138 people in Joplin, Missouri. On May 24, a further outbreak in Arkansas, Kansas and Oklahoma killed 16 more people.[29]

The total number of these supposedly rare EF5 tornadoes *doubled* in 2011. For many Americans, it was this devastation that finally nailed down the idea that we are, indeed, facing new climatic phenomenon.

Despite these overwhelmingly convincing events, though, both evangelists and scientists were unwilling to claim any connection between climate change and the gathering intensity, frequency, duration, and power of the new-fangled tornadoes of the 21st century. Admitting personal failings or dogmatic fallibility is understandably dangerous for church leaders because it leaves them open to challenges about the genuineness of the belief system and their authority to lead, and this intensifies their denial. But there seems to be a moment when denial becomes suspect, and recently evangelical American Christians have reached this point. As a consequence, substantial attrition is now afflicting America's evangelical churches.

But what about the scientists?

Few ever claimed that global warming was false, but nearly all claimed that no definitive connection had been established between the increases in extreme meteorological events and climate change. Those who put their faith in scientific research sometimes hoped that because the issue was so vital to the human survival, some rash climatologist would put caution to the side and admit there is a strong possibility (perhaps a likelihood) that tornadoes were and are intensifying and that climate change is by far the most likely culprit. But only a few scientists were willing to take such an extreme step. Back in 1949, a few months before he died, conservationist Aldo Leopold, the father of environmentalism, explained why our scientists refuse to lead:

> There are men charged with examining the construction of the plants, animals and soils which are the instruments of the great orchestra. These men are called professors. Each selects an instrument and spends his life taking it apart and describing its strings and sounding boards. This process of dismemberment is called research. The place for dismemberment is a university.... All [professors] are constrained by an ironbound taboo which decrees...[that] the detection of harmony is the domain of the poets.
>
> Science contributes moral as well as material blessings to the world. Its great moral contribution is objectivity, or the scientific point of view. This means doubting everything except facts. It means hewing to the facts, let the chips fall where they may.[30]

Between the years 2008 and 2012, Americans were confronted by the frustrating lack of spiritual and scientific leadership, so they began to do what they often do anyway: they made up their own minds. Increasingly, stories about the impact of climate change appeared in the popular press, and these were read avidly by Americans perplexed by the rash of extreme meteorological phenomenon: tornadoes, hurricanes, rains, floods, blizzards, heat waves, and the perpetual drought then (and still) scorching the American South. [31]

(I write this on Thursday morning, January 31st, 2013. Yesterday and today, a cluster of 13 *winter* tornadoes again ripped through Alabama, Arkansas, Georgia, Illinois, Indiana, Mississippi, Missouri, Oklahoma and Tennessee.)

Superstorm Sandy

> *Long as I remember the rain been comin' down*
> *Clouds of mystery pourin' confusion on the ground.*
> *Good men through the ages tryin' to find the sun.*
> *And I wonder still I wonder who'll stop the rain.*

JOHN FOGERTY, CREEDENCE CLEARWATER REVIVAL,
"WHO'LL STOP THE RAIN?" (1970)

People do not migrate *just because* of tornadoes, any more than Californians leave their homes on account of the occasional earthquake. Nonetheless, tornadoes contribute to the massive ongoing damage caused annually by climate change. At a time when the American economy is challenged to borrow money to pay interest on the nation's debt, the cost of reparations forced on us every year by an increasing number of extreme weather events eats hungrily into local, state and national resources. The persistence of this problem forces hard choices between repairs and economic necessities like social services (police, fire, water, hospitals and schools). Even though in Michigan the culprit was recession and the mortgage-housing crisis, *not* climate change, Detroit's deserted streets and houses are nonetheless a stark warning of what's to come. Once America's fourth largest city, Detroit's population plunged 25% in the last decade to a little over 700,000 people. Pulitzer laureate and native-son Charlie LeDuff puts it most eloquently: "It is awful here, there is no other way to say it.... I believe Detroit is America's city. It was the vanguard of our way up, just as it is the vanguard of our way down."[32] When repair becomes unaffordable for whatever reason, the infrastructure of our cities goes into decline. In Detroit, simultaneously, the crime and murder rates are soaring, so it is now reputed to be the most dangerous city in the United States. As bad as this is, the death-rate during future meteorological disasters (heat waves, for example) will skyrocket among those who remain in place. Mostly, it is the old and infirm that stay, as the young and healthy leave for greener pastures.

This spiral-of-decline is called "collapse." It's already painfully familiar in the rest of the world, but it's a relatively new event in the United States. Nonetheless, Dmitry Orlov claims it is becoming the most prominent and quickly emerging "taboo topic" among upper echelon professionals:

> For certain specialists—scientists, engineers and, more recently, those working in finance—collapse is fast becoming the unmentionable elephant in the room, and the enforced silence causes them much frustration, since it is becoming increasingly difficult for them, based on the data at their disposal, to formulate, in their own minds, a scenario that does not culminate in collapse.[33]

During the recent recession, many municipalities declared bankruptcy and put an increasing burden on their state economies. And while municipalities are corporations, which can declare bankruptcy, individual states cannot. In some states, the burden of an economy broken by the lack of local growth is too great. Outmigration is the result. This process is accelerated by climate change, which now adds to the list of factors that produce regional economic disasters. The most pronounced impacts affect three of the most populous regions of the United States: California, the Southwest, and Greater New York.

In addition to the immediate cost of reparations, with each successive disaster, insurance premiums ramp up—just as they do after repetitive car collisions, no matter who is "at fault." As insurance gets more expensive, it is simultaneously becoming more intricate, complicated, and qualified. Less insurance is getting more and more expensive. Moreover, it is increasingly difficult to acquire. Because of this, areas prone to extreme weather events are left doubly vulnerable; as each event damages the local economy more severely, the affordability of protection and preparedness declines. At such times, smart money moves, just as businesses are now leaving California. This slow spiral can be temporarily arrested by clever, experienced, determined leaders like Jerry Brown, but climate change and the declines in infrastructure and in quality of life cannot now be stopped by a single political leader. In fact, it's safe to say at this point that it simply cannot be stopped at all. Like it or not, we are entering an era of consequences.

Good CEOs, who understand the larger patterns of things, understand this unpleasant truth. Abstract thinkers can now see very clearly that the global economic growth-spurt generated in 1945 by the *Pax Americana* is winding down. It is time to go elsewhere; it's time also that we acclimate ourselves to much more modest standards of living. "We are vulnerable," Governor Cuomo says categorically. "Anyone who thinks there is not a dramatic change in the weather patterns is denying reality. We have a new reality and old systems." [34] Cuomo's bailiwick, the Greater New York (and TriState) region, has a population of 30 million people, all of whom are living at sea level, very close to the Atlantic shore. Nearby, the Baltimore-Washington Metroplex is home to another 9 million coastal residents; Philadelphia adds 1.5 million to the number of citizens exposed to extreme weather events like Superstorm Sandy. Cumulatively, nearly 20% of the population of the United States (more than 60 million people) lay in Sandy's path.

These same people will be victimized again and again in coming years as the North Atlantic's increasing temperatures drive larger and larger storm surges into populated areas. As Climate Central (a group devoted to researching and reporting the science and impacts of climate change) puts it:

Most people think the greatest danger from a hurricane lies in the storm's punishing winds and torrential rains. But those who live on hurricane-prone shores…know there's something else to worry about. It's the storm surge, a huge pulse of seawater pushed up onshore by the coming hurricane, like a slow-moving tsunami.… Much of the devastation caused by Hurricane Katrina in 2005…came when the surge of Gulf water overwhelmed the protective levees to flood the city.[35]

In the TriState region, the danger is enormous and persistent. And if only 1% of the residents of Greater New York decided to leave the region, that would still represent a greater movement of people than the estimated half-million Americans who came to Canada during the Vietnam conflict. In terms of contemporary American demography, it's a barely noticeable number.[36] But in Canadian terms, 600,000 people is an immense figure. (Annual immigration to Canada has never been that high.) It is a question of dropping the same-sized rock into very different-sized ponds. In the small pool north of the 49th parallel, the impact will be huge although, of course, not all Americans fleeing climate change will become north-bound migrants. Many, at first, will look for more comfortable options by relocating elsewhere in the United States. In these pages, I am trying to demonstrate how the number of such options will shrink increasingly with the passage of time.

The original losses caused by Sandy were estimated at $50 billion (six months later, the estimates now hover around $65 billion, but the final figure will probably be higher). Sandy began as an unseasonable North Atlantic hurricane one week before Election Day in November 2012. Just as it is with drought and tornadoes, the decisive factor in creating a hurricane is heat. Nowadays, there is much more of it. Since 1750, our consumption of coal and gas has poured 350 billion metric tons of carbon dioxide into the atmosphere. This has raised the level of CO_2 particles in our atmosphere from 280 to 390 ppm.[37] The result is a hotter earth and a hotter sea surface. Just before Sandy formed, "sea surface temperatures...were 5°F above the 30-year average, or 'normal,' for [the] time of year over a 500-mile swath of the coast."[38]

Warm water creates and feeds hurricanes. Such storms used to be once-a-century events, but a central point of global warming theory is that North Atlantic hurricanes will increase in the coming decades as the world heats up and the tropics expand toward the poles. Meteorologists now record more storms at the end and after the traditional hurricane season: "Normally there are 11 named Atlantic storms. The past two years have seen 19 and 28 named storms. This year [2012], with one month to go, there are 19."[39]

Climate scientist Katharine Hayhoe of Texas Tech University is quite precise about the historical difference in sea temperatures: "The Atlantic Ocean...is about 2° warmer on average than a century ago," she says.[40] Another scientist, Peter Gleick, points out that one immediate impact of this heat is sea level rise. The oceans, he says

> are rising for two reasons: thermal expansion and an increase in the volume of the ocean. As the planet warms, so do the oceans, and warmer water takes up more space than colder water. As ice in glaciers and the ice caps on Antarctica and Greenland melt, the volume of the ocean grows.[41]

Sea level rise, of course, is a vital issue in considering the damage that Sandy caused and projecting what damage future North Atlantic superstorms will have in a region as vital to the global economy as New York City. The higher the sea level, the worse a storm surge's impact will be during a hurricane, as meteorologist John Abram explains: "Most people think they don't have to worry about...a few millimeters...but every inch we get makes a storm surge worse."[42]

Emphasis on the destructive power of the storm surge itself has led to new methods of describing the force of hurricanes. The

Saffir-Simpson scale (SS) "currently used to communicate the disaster potential of hurricanes…is a poor measure…because it depends only on intensity."[43] Recently, meteorologists have developed a more subtle means to explain how and why a storm like Katrina—which in 2005 was described as a mid-level Category 3 storm on the SS scale—carried with it such enormous destructive potential. The new method scientists have developed is more descriptive of the power of superstorms like Sandy.

Instead of wind speed, the new storm scale measures the Integrated Kinetic Force (IKE) of major storms. In Sandy's case, because the front was 1,000 miles across, it is easy to imagine that even if the storm were traveling slowly, it would have a huge amount of bulldozer-type energy, pushing into and moving anything in its path. This is the kind of power that drives a storm surge when it strikes a coastline and allows it to erase built structures even before the wind arrives. This is exactly what happened with Sandy. Researchers describe Sandy's IKE rating as second only to that of Hurricane Isabel (2003), and "higher than devastating storms like Hurricane Katrina, Andrew and Hugo."[44] The IKE metric can be used as a precise measure of a major storm's force. By this scale, Sandy had a force of 140 terajoules. One terajoule is equal to a trillion joules, or about 277,778 kilowatt hours. This means Sandy contained "more than twice the energy of the Hiroshima atomic bomb."[45] William Westhoven, the former performing arts critic for the *Daily Record*, provides us with an eyewitness account of the storm as it struck the Jersey shore:

> Sandy blew with the vengeance of a jilted lover, her blind fury growing impatient throughout the night as though she were popping transformers like five-hour energy drinks. Her path of destruction was as indiscriminate as a rubber-suited Japanese movie-monster, leaving some heels and toes of the neighborhood flattened like a footprint while [other] nearby homes were completely spared.[46]

East Coast Water Levels

The geophysical changes that global warming brings to America's East Coast are unique. In addition to the increased volume and thermal expansion of the oceans generally, New York and the northeastern coastline are impacted by the rapidity of Greenland's melting ice cap. In the past 20 years, 4 trillion tons of Greenland's ice have melted into the Atlantic and increased the volume of the world's oceans.[47] And another 344 billion tons of ice is melting annually in Greenland and Antarctica.[48] The average global increase in sea levels is about 8 inches, but on America's Atlantic coast the figure is higher: "Water levels around New York are nearly a foot higher than they were 100 years ago," said Penn State University climate scientist Michael Mann.[49] Scientists speculate that as cold water from the Greenland ice-melt sinks and flows south, it pushes up the warmer coastal waters, creating a sea-level-rise hotspot along an 800 mile stretch of the American coast in its most densely populated area: from Cape Hatteras, North Carolina, to Boston.[50]

The area of this sea-level-rise hotspot is the area most vulnerable to the North Atlantic Superstorms of the 21st century, but there is more bad news. The rate of Greenland's ice-melt is accelerating, so sea levels along the East Coast will continue to rise more quickly than in other parts of the world.[51] A NASA scientist points out that these changes will not follow regular rates of change. As we have seen, because change cancels probabilistic "stationarity," all bets about future events are off. There is no accurate way to predict anymore:

> It's really shortsighted to assume that the next hundred years of sea level rise are going to be like the last hundred years.... We're already seeing glaciers and ice sheets melt more quickly, and the ocean absorbing more heat and expanding—things that drive sea level rise.[52]

Moreover, as the water of a storm surge is forced in front of the winds, it becomes channeled and directed in a pattern called *radiation stress*.

In the past, this factor has played a role in intensifying the storm surge of hurricanes like Katrina. Radiation stress depends on the unique geophysical features of wherever a storm makes landfall.[53] Even without the radiation stress characteristic of the southeastern coast, however, many predict that the northeast Atlantic coast's storm surges will "cause 20% more damage by the year 2030."[54]

Twenty percent more damage is truly frightening, considering the $65 billion that Sandy is known to have caused. But there *were* special conditions that made Superstorm Sandy's timing particularly vicious. An especially high tide (sometimes called a *king tide*) in the TriState region raised water levels to their maximum just as a rapid conveyor was bringing half of the superstorm into the region so quickly that it was able to combine with another storm. Together, these and other factors raised the storm surge that struck lower Manhattan and the Jersey Coast to a height of 13 feet. The devastation of that wave was comparable to a tsunami, and it flattened most of the low-lying communities from Staten Island to Far Rockaway. Sandy's intensity exceeded the predictions of climate change models by about 70 years, according to Art De Gaetano, head of the Northeast Regional Climate Center (NRCC) at Cornell University:

> The impacts we saw [during Sandy] were actually impacts we did not think we would see until the 2080s, such as the flooding of lower Manhattan, and in Long Island, the vulnerability of subway tubes and some of the airports.[55]

Renowned climatologist, Kevin Trenberth, explains how much of Sandy's raw power was owing to climate change:

> Storms typically reach out and grab moisture from a region 3 to 5 times the rainfall radius of the storm itself, allowing it to make such prodigious amounts of rain. The sea surface temperatures just before [Sandy]…were…5°F above the 30-year average, or "normal," for this time of year over a 500-mile swath off the coastline from the Carolinas to Canada, and

1°F of this is very likely a direct result of global warming. With every degree F rise in temperatures, the atmosphere can hold 4% more moisture. Thus, Sandy was able to pull in more moisture, fueling a stronger storm and magnifying the amount of rainfall by as much as 5 to 10% compared with conditions more than 40 years ago.[56]

The relationship between climate change and Sandy is significant because it explains the amount of damage inflicted on America's pivotal economic region. It should also send chills down the spine of anyone who realizes the implications of such intensifying, more frequent and more unseasonable North Atlantic superstorms. Greenland's ice-melt is increasing at an alarming rate, which makes predictions of future superstorm occurrences and likelihoods suspect and of little real value. New York is preparing itself for sea level rise of about 4.5 feet by the end of the century; such adaptation is formidably expensive, but it may also be quite useless. What we know will happen is that Manhattan, in fact, *all* of the TriState region, will continue to be battered—increasingly—by superstorms like Sandy. This will have an enormous impact on health, infrastructure and the economy that will cause downsizing and outmigration.

Canadians or Alaskans with foresight might attract potentially mobile businesses ready to flee the sites of future climate-change disasters by creating conditions favorable to relocation. But the time to do this is right now. In any case, very soon, Americans with youth, energy and foresight will begin to ask themselves why they need to stay in New York City.

And when they ask that question, they will also ask: *"Where is the best place to go?"*

CHAPTER 8

The Safest Place to Go

Only the exquisite work contingent known as "Canada,"
which cleaned the cattle cars of the incoming trains, could rival
ours in terms of the food that could be found in the garbage.
We called it "Canada" because in our minds that word
signified a land of great abundance.... Alas, this
"nutritious" period of our captivity was very short-lived.

SAMUEL PISAR, OF BLOOD AND HOPE (1980)

Faster Than Expected

No original estimate of a specific global warming impact has ever been overstated. Generally, these predictions begin modestly before revisions appear that accommodate the severity and speed of climate changes as they actually take place. Earlier chapters demonstrate a common movement from caution to crisis in drought predictions for the American Southwest, in predictions of the deterioration of the cryosphere, in predictions of the occurrence of North Atlantic hurricanes, in predictions of sea level rise and—unfortunately—in predictions of species migration and extinction. Until now, I've withheld listing these successive examples in order to reserve making this point: only as the processes of climate change unfold do the intricacies of feedback mechanisms become clear. The increasing transparency of these mechanisms has always, so far, been accompanied by the increasing severity and urgency of specific climate changes including drought, flooding, heat waves, hurricanes and ice-melt. You could say that, as our knowledge increases, the fog of our ignorance lifts and the horizon of humanity's true predicament expands.

Our continuing surprise at the rapidity of various climate changes *and* their effects explains my departure from the sluggish pace of scientific prediction. Like a basketball player, I'm trying to move ahead to where the ball *will* be because the stakes are high, and my concern is not with verifiable certainty, but with survivable likelihood. I respect the dedication, training and integrity of people-of-science, but I prefer to follow my soldier-father's example of preparing for the worst so that whatever happens, it will come as a more welcome surprise.

This sentiment is shared by only one climate scientist I know. He has committed himself completely to his minority opinion by buying an expensive tract of land in northern Canada where there is supposed to be "ample water." A child of Holocaust survivors, he feels a palpable obligation to his forebears to provide his own children with "a backdoor" on the unfolding history of our current century. Describing himself simply as "a link" in a very strong intergenerational chain, he does not want his name publicized. This has more to do with maintaining sensible security than with fear of mockery. He is not a timid man.

How Bad Will it Get?

A key issue throughout North America will be water—a topic that takes a profoundly historical dimension in Canada, which began as a country explored by intrepid, self-sufficient voyageurs. When they returned from the furthest reaches of the continent, their large canoes rode deep in the water because they were laden with 2–4 tons of animal pelts. The exploration of America, on the other hand, depended on stalwart walking figures like Lewis and Clark or Daniel Boone who, like Tecumseh's soldiers, disappeared again and again into the dense foliage of the Eastern woods armed with little more than a flintlock, a tomahawk, and a "can-do" attitude. One of the United States' best poets recognized Boone's achievement and wove it into his myth of America:

> By instinct and from the first Boone had run past the difficulties encountered by his fellows in making the New World their own.... [He] lived to enjoy ecstasy through his single devotion to the wilderness with which he was surrounded. The beauty of a lavish, primitive embrace in savage, wild beast and forest rising above the cramped life about him possessed him wholly.[1]

Canadians, a more stolid people, have been less preoccupied with mythmaking. They did not valorize explorers like Davy Crockett or Daniel Boone. It was an anonymous navy who heard the same seductive call to exploration that stirred the 12-year-old Boone during his early visits to Indian villages on the Pennsylvania frontier. American have *The Leatherstocking Tales*, but the unnamed force that tugs people away from civilization rarely finds a voice in Canadian letters. Those who feel it most strongly are often society's least articulate members, busily preoccupied with practical, daily affairs in their challenging lives. Even so, the urge to shuck the complexity of civilized life for parts unseen has been part of the human genetic fabric since first leaving Africa. It plays a vital role in all migrations to pristine, new lands, although its expression rarely finds its way into print and when it does, its vocabulary may leave splinters under your skin:

> I been in the country trappin' every winter since I were twelve.... Out of the last 30 summers I only missed three in the bush.... There's nothing for me like travellin' a new river, seem' the beeg hills close in behind and the bends open up ahead, breathin' cold sunshine and seem' a new country every day.[2]

The voyageurs reached as far inland as Great Bear Lake in the Northwest Territories—discovering, exploring and naming innumerable Canadian rivers along the way. Some 300 years later,

Canadians still take the abundance and ubiquity of their freshwater for granted. Undoubtedly this bounty will be the country's greatest blessing in the coming century. In select northern regions, precipitation will actually *increase* by 40%; these are the choice destinations to which migration should be directed (but more on that later).

Southern Canada, on the other hand, is becoming drier as of yesterday. Many regions will burn in uncontainable fires whose size and scope will stagger the imagination and bring on epidemics of respiratory illnesses. Rivers will boil. Sadly, there's little on the foreseeable horizon to prevent this. Fully half of Canada is covered by forest. The Canadian woodland comprises 10% of the world's tree cover. Despite contemporary statements about what we can and will do to save the boreal forest, little is being done. There is no coordinated effort to study and eradicate the beetle species that will reduce our forests to kindling and prepare them to burn. Meanwhile, rising temperatures have made wildfires throughout North America larger, more frequent, and much less manageable in the past 20 years. For this reason, I believe—reluctantly—that only a fraction of Canada's vast timberland will survive the coming decades of drought, year-round insect infestations, and megafires. As the tree line creeps northward from 68°N (its highest point in Canada), forests in the warming South will dehydrate from drought, succumb to insect infestation, and burn irretrievably. This process has already begun. The "carbon summer" is coming.

Still, even in these early decades things will be much better in Canada than in many areas of the United States. So to Canada they will go. As the war of 1812 ended, Thomas Jefferson predicted that "the acquisition of Canada...will be a mere matter of marching." Jefferson was wrong, but within the span of this current generation (before 2030), people will begin to leave the Southwest and (especially) California due to a complex of environmental and economic problems that have at their center an insufficient water supply. Soon, already diminished supplies will reach a point where they can support neither agribusiness nor the second most concentrated population in the United States. People will leave, not because they can't survive where they are, but because of the fact that life is getting worse. It will become obvious, and there will be much better places to go. During the next few decades, freshwater in the Southwest will diminish by at least 30% more as temperatures increase and the Ogallala Aquifer toxifies; shortages have already resulted in droughts in 13 states; in the summer of 2013, that number will probably double. In California, ongoing drought is now complicated by year-round occurrences of forest megafires that add emotional and economic stress to a state so economically challenged it can no longer enforce its own prison sentences or pay its own teachers.

I spent 15 good years teaching college in L.A.'s ghettoes while enjoying the grace of Californian hospitality. My mind automatically compares every sunset to that of my first night watching the sun sink past Santa Monica. Behind me, a couple of indeterminate gender bustled in a grey-and-red striped, flannel beach-bag while a big black man on sax guarded and serenaded them. I sat in the sand with a bottle of Anchor Steam between my legs and scholarship papers in my pocket, feeling like I had just won the lottery. A week or two later, I met my wife at USC, and in 1989, we were married in the same Botanical Gardens in Santa Barbara fated to burn in a forest fire in May, 2009. My eldest son is a California native and avid Lakers fan. We revisit the state whenever we can. So, these days, I'm saddened when I surf into the *Los Angeles Times* or *San Francisco Chronicle*. The spiral of economic decline that afflicts Californians is accompanied by increasing environmental challenges. I'm lucky enough to have two homes, but I'm preparing to lose one of them. In California, social collapse is imminent, and water is its deepest cause. The question now is "Where will all those Californians go?"

As people abandon the Golden State, more and more thirsty people from Arizona, Colorado, New Mexico, Texas and Utah will also conclude that the climate is irrevocably changing and that the further north they move, the better their chances will be over the long term. Those with sufficient foresight and the luxury of economic mobility will fan out early into the regions surrounding the northern cities of Portland, Seattle, Spokane, Helena, Billings,

Duluth and St. Paul. The bravest, most informed, and most affluent Southwestern residents will leave the lower 48 emulating the north-bound colonists of the New Deal who settled the Matanuska Valley between Palmer and Wasilla, Alaska. But most people fleeing the early manifestations of climate change will at first resist making such a fearful decision on account of distance, climate, or widely different social amenities. The people who stay in what I think of as North America's "midriff" (the waistline of northern states along the Canadian border), will eventually become so numerous that municipal infrastructures will stress to the breaking point, and state budgets will fail. At that point, further human movement will become necessary.

For several decades, migration will instigate further migration. The question *"Where is the safest destination?"* will be at the forefront of most Americans' minds.

Climate Change in Canada

It may surprise my American readers to learn that Canada is even more urban than the United States. Eighty-five percent of Canadians live in cities (compared to 80% in the United States). Our most famous literary critic, Northrop Frye, once wrote that Canadians have a "garrison mentality." He meant we huddle in urban centers and shun wilderness, the enormity of which frightens us either because *we* are so timid or because *it* is so overwhelming. Margaret Atwood, Canada's most famous novelist, is more charitable to Canadians, writing simply that they are preoccupied with survival, and, once it is achieved, we like to put our feet up. Like Atwood, I was born in Ottawa, where the forest of the river valley and the nearby Gatineau Hills turn into a spectacular gold-red pageant every fall. Its beauty can overwhelm you, and our winter is a formidable force, but the forests of the eastern border region harbor no insurmountable obstacle to human survival. Yet.

Soon, however, droughts and heat waves will envelop most communities in southern Ontario and Quebec. Enormous fires will then develop that will consume the deciduous forests of my youth, along with those of upstate New York and Vermont. Rising temperatures will increase evaporation so much that even the largest lakes will subside at least 5 feet by mid-century. In the Midwest, water levels in the Great Lakes and their surrounding aquifers are already falling. As lake levels subside, Huron and Erie will toxify and fill with algae forests, losing their remaining fish. Along the southern shore of Lake Michigan, fires, heat waves and declining freshwater will force many residents in the Chicagoland metroplex (from Milwaukee to South Bend) to relocate. To residents of border-states threatened by climate change, the proximity, abundance of freshwater and lack of population density in Canada will make migration northward look easily attainable and very inviting. Small, privately owned watercrafts will make the northward migration of American lakeshore residents as unstoppable as southbound smuggling was for my great uncles during Prohibition. For American residents of the Great Lakes, escape to Canada will be as close as the family canoe or motorboat. But before families push off from the shore, concerned parents will huddle over maps and talk in hushed, worried tones around the kitchen table trying to answer the question "*Where is the safest place to go?*"

I don't honestly know when residents of the eastern states will begin their journey north. In 1995, the IPCC estimated there would be a maximum of 58 centimeters (22.8 inches) of rising seawater by 2100.[3] Recent research has amended this prediction to between 1 and 1.5 meters (about 40–60 inches), but Ban Ki-moon, secretary general of the UN has said that we should expect a minimum of 2 meters of sea level rise by 2100.[4] What's more, his estimate is a bit conservative. The seas will not rise uniformly, and the proximity of Greenland will likely cause greater sea level rise on the East Coast before ocean levels equalize. In California, the Pacific Institute now predicts a disastrous minimum of 1.4 meters (55 inches) of sea level rise throughout California by 2100. Exactly how much it will be depends on the timing of the ice-melt in Greenland and the Antarctic. All we can really say for sure is that the rise is happening faster than predicted. Intuitively, I believe we have not yet seen a correct upward estimate in print. In conversation, however, some scientists admit a rise of "several" meters is likely by 2100 (several, of course, means more than two).

To me, this indicates that at some point well before 2100, life for the 164,000,000 Americans who live within 50 miles of the ocean will become unendurable. Storm surges and sea level rise will erode the shoreline and compromise the infrastructure and freshwater supply of coastal cities, making internal transportation impossible—while also incubating very nasty water-borne illnesses. This will be especially true for those 60 million Americans who currently live in the TriState area. Of course, some coastal dwellers will simply move upslope temporarily, but this is not a strategy available for residents of Manhattan who, except for the mild incline of Sugar Hill in Harlem, possess very little in the way of "upslope" since the "leveling hand of improvement" was generously applied to New York City to realize the urban "grid" as first conceived in 1811.[5]

In addition to New Yorkers, many other people displaced by rising water and increasing storm surges will see clear consequences for their immediate future. The majority will begin to move inland, traveling west and north searching for more livable sites. For reasons of proximity, many will probably settle well above the disappearing Great Lakes. The teeming flora and fauna of the lush, wet boreal forest will beckon, as seemingly safer places to live. But some people will realize that this too will eventually be an unsafe region—one subject to forest fires as climate change dehydrates the ground and the canopy. These people will again be deeply troubled by the question "Where is the safest place to go?"

Back among the earliest "adapters" on the West coast, there will be a small percentage of Americans with some connection to Canada: it might be a branch of a family that originated north of the 49th or maybe a new branch begun by a draft dodger during the Vietnam era; perhaps people will feel connected to their northern neighbor through a college-aged child who took advantage of Canada's inexpensive post-secondary education, some former college friend, a business associate, or simply the pleasant memory of a favorite vacation spot like a ski resort or summer cottage on a pristine northern lake.

To Americans with any of these connections, Canada will be a known quantity, so they'll be quicker to embrace the existing option of dual-citizenship. Without wishing to strain my credibility or my reader's credulity, I *imagine* that like the exo-dusters or the so-called "wetbacks," many of the first migrants will be single, unattached people with little to hold them to the worsening conditions of life in the United States. They will also possess sufficient energy, youth and enthusiasm to jump at a chance, however slim, for a better life. They will be ready, willing and able to adapt and will have (as was once said of the 70,000 Loyalists who left the United States after the Revolutionary War) "resources in themselves which other people are usually strangers to."[6] Canada will benefit from their presence, as it did from the roughly 10,000 fugitive slaves who arrived in the mid-19th century, or from the estimated 100,000 or so Americans who moved north during the era of the Vietnam War.[7] The best immigrants will be those who do not look back, but instead decide early to remake themselves and to construct solid, new lives in their new home.

This is always easiest for the young.

Soon, these early adapters will be joined by slightly older couples with better economic mobility looking for alternative ways to provide for the well-being of their children. Like the younger first wave, they will probably succeed in entering Canada legally, before national immigration laws change in a vain attempt to slow the inevitable flood of migration from south to north. Recently, Sir Nicholas Stern, a renowned international authority on the economic aspects of climate change, issued this warning:

> Let us be clear, these dangers are of a magnitude that could cause not only disruption and hardship, but mass migration and thus conflict on a global scale.[8]

After such migration and conflict begin, it seems likely that Canada's naturalization laws will begin to more openly reflect the country's persistent undercurrent of anti-Americanism. Winston Churchill famously described Canada and the United States as more "mixed up together" than any other two nations. But despite the fact that the history of Canada and the United States is intertwined and comprises, really, one larger history—the history of European and African migration across a purloined continent—still, Canadians have defined themselves for nearly a century and a half simply as "non-Americans." We have no overarching myth of self-definition, as our powerful neighbors do, and this, I believe, is really the deepest source of our abiding resentment of the United States. In any case, the "lifeboat ethics" described in Chapter 1 will begin to surface just as soon as the new immigrants strain the economic resilience of their adopted home. There will be significant Canadian resistance to any noticeable migration northward, and this resistance will grow into profound social conflict that will trouble Canadian society for a generation or more. Yankees will become the new Okies. However, geopolitical realities will abruptly change once climate change makes the United States poorer and Canada richer in terms of the only currency that matters: freshwater. Since the entire continent of North America now contains 460 million people, and Canadians represent 7.5% of that total, it's very hard to see how Canada could resist the will of its powerful neighbors when they become increasingly hungry, thirsty and desperate.

These events will decide Canada's future as a nation. All of North America's governments should begin planning for that now. Robert McLemen, a Canadian professor of human geography, made the point in 2007 that any increased migration from the United States will have a profound impact on Canada:

> If one-tenth of one percent of Americans got the idea of moving to Canada, it would represent 300,000 additional migrants, well beyond the current worldwide total of annual immigration to Canada. Two-tenths of one percent makes 600,000.... Even a tiny, fractional change in American migration patterns could bring enough migrants to challenge Canada's ability to accommodate them economically and in-

corporate them into our social programs. Rapid population growth in regions of increasing climate risks will provide the raw ingredients for [a] large-scale American migration.[9]

Where Not to Go

Despite Canada's international reputation for natural abundance and safe neighborhoods, simply crossing the border into the North will not end anyone's vulnerability to climate change. Global warming has already reached into every corner of North America. Many state birds can no longer be found in their official ranges; and the "hardiness zone maps" of the US Department of Agriculture are now functionally obsolete because they're based on climate data collected before 1986. The southern 48 states have been bombarded by ice storms, a formidable number of twisters, and disastrous spring floods just in the last few years. But in Canada, too, meteorological change has become eerily commonplace. These changes offer an early taste of what we will all face in our various localities during the coming century.

Drought

In Canada, the most obvious place to avoid is the southern prairie—especially the large, semi-arid region known as *John Palliser's Triangle*. Stretching from British Columbia's eastern border across Alberta and Saskatchewan into western Manitoba, this immense wedge-shaped region already experiences the panoply of problems I've described in this book. *Empire of Dust*, a regional literary treasure, describes the formidable challenges that turn-of-the-century settlers experienced here—long before the Dust Bowl or global warming arrived to make matters worse:

> The sun could bake crops to a crisp in less than ten days, and it could knock a man out from heat stroke in a few hours.... Desiccating winds scorched and parched and burnt.... It could rain for days in the spring, sometimes a quarter or a third of the year's allotment, filling the sloughs, turning the

roads into bogs and the fields into lakes, with the whole prairie awash, but it always took the winds less time to dry it.[10]

These days, life in the southern prairie is considerably worse. The main difference is an absence of summer snowmelt since "drought and heat consume...snow waters" (Job 24:19). During the worst years of the Dust Bowl, winter could be counted on to bring sufficient snow cover to buffer the hottest weather of June, July and August. This has changed. Wintertime precipitation over the Canadian prairies has actually increased since the 1980s, but winter is now shorter. Spring comes sooner, and, even at the highest elevations in the foothills of western Alberta, the weather is hotter and drier, meaning that snow cover turns to ice-melt much earlier in the year. In high summer, prairie winds sometimes blow alkali dust out of dried, shallow lakebeds, causing local epidemics of respiratory illnesses.

Also during the past decade, Alberta's Saskatchewan Glacier, source of the South Saskatchewan, Bow, and Red Deer Rivers, has receded rapidly. Once an easily accessed and well-photographed tourist attraction, it no longer waits patiently beyond the shoulder of the Banff-to-Jasper highway. In 2002, a local journalist who hiked two miles to reach the glacier put it this way:

> The recession of the ice has been so fast that the forest has not had time to re-vegetate.... Huge gravel moraines [are on] either side of where the ice used to be.[11]

Early ice-melt has serious implications for Alberta, for Canada, and for the world. Summer droughts kill water-dependent oilseed crops like canola, but they also seriously damage traditional dryland crops like wheat and sorghum. In 2003, 81% of Alberta's wheat was lost to a devastating combination of drought and grasshoppers. Each successive drought reduces surface soil moisture, making it impossible to plant in the coming spring. "Severe" moisture limitations had once been predicted for the year 2040, but they are already happening. For this reason, the southern prairie is experiencing the same outmigration that Frank and Deborah Popper once described in America's High Plains. Grain farmers like Keith Anderson are moving north, abandoning the southern Alberta wheat farm his American grandfather pioneered 100 years ago by relocating to a cattle ranch north of Edmonton.[12]

The Andersons may not be safe even in northern Alberta. Although their new ranch outside Rochester is currently surrounded by lush vegetation, most of Alberta experienced killing droughts during the first decade of the new century. In recent years, northern Alberta lost most of its forage crops to drought; this puts the already thirsty livestock and cattle industries into double jeopardy. In Alberta, only a narrow corridor along the Saskatchewan border in the northeastern part of the province has been immune to drought so far. The wettest part of Alberta, from Wainwright to Cold Lake, lies along the Athabasca River which feeds Saskatchewan's Lake Athabasca, the largest open body of freshwater in a vast system of rivers, lakes and wetlands that characterizes Saskatchewan's inviting Athabasca Plain above the 59th parallel.[13]

Some suggest that two regions (the boreal shield and boreal plain eco-zones) south of Lake Athabasca might become the heart of a future breadbasket as climate change stretches northward and warms what is now 23 million hectares of forest. Along the southernmost margins of the boreal plain, more than 10% of the land is currently devoted to raising livestock. No one, however, has systematically tested the soil in either region to see if large-scale grain farming might be feasible.

And then there are the trees themselves: the boreal forest is currently an enormous carbon sink that helps replenish the planet's oxygen. Allowing this timberland to disappear through megafires or farming means a substantial reduction in the efficacy of our planet's breathing apparatus. We should tread very lightly here.

South, lies the Saskatchewan River, whose summer levels have dropped in recent years due to early snowmelt. In 1738, at the age of 53, one of the few voyageurs whose name we still know—Pierre Gaultier de Varennes, Sieur de la Verendrye—reached (and some say discovered) the Saskatchewan River while searching for a route westward into new lands untapped by the Hudson's Bay Company. Verendrye had been wounded nine times in battle during a military service that began when he was 12 years old. But the brutality of the winter of 1738 chilled him deep into his aging bones: "Never in my life," he later wrote "did I endure so much misery, pain and fatigue as in that journey."[14]

Today, winters are much milder along the river's shores, and its water level is also substantially lower. For this reason, the prairie and much of Canada's grain harvest dies regularly in the same superheated summer droughts that trouble all of southern Alberta, Saskatchewan and Manitoba. Across the country, these same droughts bring on the megafires that have been raging for over a decade. Very few areas of Canada will be safe from such fires, but some are already much more dangerous than others.

Megafires

In northern Quebec along the eastern coast of Hudson and James Bays, the tree line begins far to the south, at 56°N, reflecting the bitterly cold impact of the Labrador Current. Since 1993, the gradual cooling of the province has been replaced by a warming trend. In northeastern Quebec there has been a marked temperature increase of about 2°C. In 2002, a late spring delayed the flow of sap, turning an entire forest into tinder when summer temperatures ramped up during the second heat wave of the new century. Lightning, of course, is a major cause of forest fires and is common in the hot, dry air of summer. In late July 2002, lightning initiated 60 fires in an area so remote that fuel constraints prevented the province's fleet of 14 water bombers from reaching the fires quickly and effectively. Unless such fires are contained in their infancy, they burn out of control until some natural force—like the arrival of winter rain—contains them. That summer, 600,000 hectares of Quebec burned, while 600 miles to the south, the cities of Baltimore, Washington, DC, and New York issued health advisories warning residents to stay out of the acrid, smoke-filled air.[15] Later that fall, Max Finkelstein and James Stone retraced the voyages of Canadian surveyor Albert P. Low, and recorded their impressions of the ravaged landscape east of James Bay:

> We are windbound on a nice beach, the site of a Cree winter camp that had been razed in a forest fire.... Most of the land we have paddled through...has been burned several times in recent decades, most recently this summer.... By evening the wind dies down and the skies clear. There is a stark beauty in the silhouette of the setting sun shining through the burnt spruce. We are struck once again by the silence and emptiness.... No birds sing, no minnows gobble up the remains of our supper, no water rings of rising trout, no ducks on the lake. A raven flies overhead, croaking, then departs. The only sound, other than the soft moan of wind through dead spruce, is the sound of our pens moving across paper.[16]

The dead area these men paddled through was as large as either half the state of Delaware or half the province of Prince Edward Island. Each summer as much as 2,000,000 hectares of forest (0.5% of the Canadian woods) now burn. Finkelstein is careful to point out that although fire played a vital role in the boreal forest during the era before climate change, the development of all-consuming megafires has changed that role:

> Fire...created the boreal forest. The survival of the forest depends on it. [It] is a force of nature that the boreal ecosystem has adapted with since the retreat of the glaciers. The cones of the ubiquitous black spruce can open and spread seed only after being baked by flames. However, with global warming predicted in our future, fires could become more frequent and burn larger areas. Already since the 1970s the amount of forest burned in Canada has increased by half.[17]

Except for urban areas and the vast, sparsely inhabited reaches of the far north,[18] the lower half of Canada is covered with trees and therefore vulnerable to the new generation of enormous fires that comprise only 1.4% of all conflagrations, but account for 93.1% of the total woodland burned in Canada every year.[19] Sharp increases in the number, size and frequency of fires are now predicted for western Alberta, southern British Columbia, and throughout Ontario.[20] Where I live, in the greater Vancouver area, health warnings against inhaling smoke have been issued often during recent summers. Despite the easterly direction of the prevailing winds, as I drink my morning cup of coffee on the back deck, I can smell the wood smoke of forests ablaze near Kelowna, B.C., 170 miles to the east. Of course, this is only a troubling hint of what's to come. In the late summer of 2010, 700 people died daily in the heat wave and forest fires that enveloped Moscow. Canadians have as yet been largely untouched by the deadly combination of heat and smoke, but this will not always be the case, and so the question persists *"Where is the safest place to go?"*

Floods

In addition to devastating fires, Canada's near future will be troubled by two kinds of flooding: riparian flooding and eustatic sea level rise. In the United States, top soil erodes ten times faster than it can be replaced.[21] Increasing spring floods along snow-fed rivers (known as riparian or riverine flooding) multiply these catastrophic losses as they did in the Midwestern United States in 2009. These floods also visited Manitoba. With temperatures rising earlier in the year, the snow melts earlier. For this reason, in 1997, a flood now called the *Flood-of-the-Century* inundated southern Manitoba, North Dakota and Minnesota and ruined 354,000 acres of farmland while forcing over 28,000 people to flee their homes in the Red River Valley. Flood damage totaled $3.5 billion dollars. The worst previous flood on record for the region occurred in 1926, but there were also large floods in 1948 and 1950.

One characteristic of global warming is that extreme events happen with increasing frequency: in 2009, the floods came again. In Canada, the floodwaters were slightly lower than in the 1997 flood, but they still affected 216,000 acres of farmland and caused about

$500 million in damage.[22] Elsewhere in Canada, flooding as a result of global warming has also increased. Spring floods have become commonplace in Manitoba, Ontario, Quebec and the Maritime Provinces. The disastrous Saguenay deluge of 1996 seems small in comparison to the extreme floods of the new century. In 2008, Phyllis Phaure was evacuated three times from her home in Peterborough, Ontario (ironically, home to the International Canoe Museum) due to flooding of the Otonabee River. There are many other vulnerable spots in Ontario, including Waterloo, Ottawa and the Cambridge-to-Brantford region.[23]

In Quebec, the town of Gaspé is now repeatedly hit by extreme floods. In 2007, Gaspé experienced its third flood of the decade. In just 12 hours, 4.5 inches of rain fell, washing away the town's three bridges and causing three rivers to overflow their banks. Tragedy struck a family in nearby Rivière-au-Renard when floodwaters broke apart the mobile home of Marie Blanchette (74) and Henri Dupuis (78). The couple's son and daughter-in-law were rescued from the wreckage, but floodwaters swept the elderly couple away. M. Dupuis's body was later recovered in the Morris River. To the dismay of her family, Mme. Blanchette's body was never found.

Touring the damage, which totaled over $10 million, Jean Charest, head of Quebec's provincial government, lamented the power and impact of the storm, laying the blame firmly at the feet of global warming: "This is not something we've ever seen before.... It doesn't compare with any natural disaster we've seen in Quebec until now."[24] Gaspé's mayor, Fran Rossy, cared very little about climate change; he knew the most relevant issue was relocation:

> If the government invests to...rebuild homes [here] it would be throwing money out the window.... People don't want to go back to those areas.... They are traumatized by this flood and they want to relocate. The compensation package that has been offered doesn't allow them to do that.[25]

In addition to riparian flooding, sea level rise (SLR) will make the coastal regions of Canada uninhabitable in the coming decades.

Bordered by three oceans, Canada has nearly 150,000 miles of coastline, the most of any nation.[26] A single meter of rising water will change the habitability of several populated regions, including 80% of the shoreline in the Maritime Provinces on Canada's Atlantic coast, as well as the coastal communities of the Mackenzie River delta on the Arctic coast and the Fraser River delta on the Pacific Coast. Long before seawater rises 1 meter, however, storm surges like the one that devastated New Orleans will severely damage the infrastructure of major port cities in Canada's Maritime Provinces, especially Charlottetown, Prince Edward Island, and Halifax, Nova Scotia.

A devastating storm surge occurred in January 2000, when a powerful winter storm blasted Canadian cities throughout the Atlantic region. Gale force winds drove waves onto the mainland throughout New Brunswick, Nova Scotia and Prince Edward Island. Charlottetown was flooded. In Pointe-du-Chêne, New Brunswick, Florence McFarlane thought it was unusual when she saw a shoe float out of her closet and cross the living room floor. She opened her front door to locate the source of the water, and a "great big wave came up on my deck, around the house and cut... the power.... I [couldn't] call for help and the water [was] coming and coming."[27] In Shediac, New Brunswick, Joe Wolkenart was trapped inside a rented cottage with his invalid wife as storm waters rose around them. Eventually they were rescued by volunteer firefighters who pulled a life raft through neck-deep water to retrieve residents trapped in their homes.[28]

Similar catastrophes will become more commonplace in Canada's coastal regions in the coming decades. Where I live, in the southern mainland of B.C., the low-lying delta farmland and my own town of Richmond lie at roughly sea level, putting over 300,000 people in immediate danger. Slightly offshore, the vast San Juan and southern Gulf Island archipelago is a unique eco-zone whose coastline and freshwater will be jeopardized by the slightest sea level rise. These islands are as rugged, beautiful and appealing as the coast at Big Sur; tragically, they will be transformed irretrievably in the next 20 years. Very soon, the residents of the southern Gulf Islands will also be forced to ask "*Where is the safest place to go?*"

Heat Waves

In southern Canada, as the tropics continue to expand poleward, the earth's four seasons will be compressed into two. There will be a wet period of intense precipitation and violent storms that will begin later and end sooner, often with record-breaking floods as the snow cover melts prematurely; then, a hot, dry season will follow. It will last well into what we now consider late fall or early winter.

In Ontario and Quebec, "Indian Summer" (my favorite time of year) will reach into November. But the droughts of the hotter, prolonged summers will leave very little tree cover to turn into the golds and reds of fall's shimmering pointillism. "May my life not be destitute of its Indian Summer," Thoreau wrote in 1851.[29] In coming decades, by July or August, declining aquifers and the regular occurrence of extreme heat waves will destroy most vegetative cover, especially the canopy of the deciduous forest, which has prodigious water requirements. Fall will just be the brown, hot period preceding winter.

Although they "are a relatively unfamiliar Canadian natural hazard,"[30] during the first decade of the new millennium, southern Canada experienced seven significant heat waves. In 2001, 2002, 2003, 2005, 2006, 2009, and 2010, extreme heat afflicted residents of British Columbia, Alberta, Saskatchewan, Manitoba, Ontario and Quebec. The coastal cities of St. John's, Newfoundland and Vancouver, B.C., have never experienced a true heat wave, but almost all other southern Canadian cities (including Victoria, B.C.) have. The longest Canadian heat waves occur in central southern British Columbia, although Saskatchewan has the record for the hottest temperature 45°C (about 113°F). In June 2005, record-breaking 40°C (107°F) temperatures in Ontario regularly caused transformers to overheat, which repeatedly shut down electrical service across southern regions of the province.[31] Far to the south, in Maryland, the same heat wave warped railroad tracks, causing a major derailment. This is a consequence of heat waves not often mentioned. In addition to their ability to kill, one of the most disastrous and costly long-term effects of "high impact heat waves" is their ability to destroy infrastructure: higher temperatures accelerate the wear on concrete, asphalt, steel, and most other common construction materials. As climate change accelerates, we will thus also simultaneously be entering a costly age of repairing and replacing North America's increasingly geriatric infrastructure.

What is truly remarkable (and frightening) is that already by 2008, a national park in the Canadian Arctic was closed after the record-breaking temperatures of a local heat wave melted the local ice and the permafrost, causing cascades so large they washed out hiking trails and forced the evacuation of tourists. Auyuittuq National Park on Baffin Island's Cumberland Peninsula is just 250 miles west of Maniitsoq, Greenland. This strait can be the most direct sea-route south from the eastern Arctic, but 19th-century whalers could only take it when they *returned* from their annual hunt because the strait was icebound until mid-August. In 1883, the ethnographer Franz Boas went to the Cumberland Peninsula to live and study among the Inuit for one year. Aboard ship between Greenland and Baffin Island on July 15th, Boas wrote to his young fiancée:

> [Dearest Marie:] Today is extremely cold, only 2°C, but I am sitting on deck as I write to you. Only today I put on my Eskimo boots, since otherwise it would be too cold on deck.[32]

By August 1, 2008—125 years later—local temperatures in July were 27°C (81°F) "well above the [modern] July average of 12°C (54°F)."[33] Uncharacteristic heat melting the local ice and permafrost as a result of climate change has enormous significance north of the 50th parallel because it changes the hydrological cycle and destabilizes the foundations of all buildings. In 1942, ignorance of the characteristics of permafrost caused the US Army to abandon and then rebuild 500 miles of the Alcan Highway, delaying its completion by one year. Civil engineer Siemon W. Muller was charged with assembling and translating everything then known about "permanently frozen ground" in order to assist the army. In 1943, he coined the word "permafrost" to describe a phenomenon originally discovered in Siberia in 1828 by Fyodor Shergin, a local businessman who was attempting to dig a well. Shergin's famous well puzzled Europe's best geological minds for the next three years. A visit to the well by German naturalist Georg Erman triggered a minor explosion of controversy in the world of natural science when he concluded there

> is a...winter temperature which prevails here in the ground at a depth where no change takes place, and even supposing that the increase of heat, from the surface to the centre of the earth, is as rapid here as in other places, yet...we would not expect to find water in the fluid state till we arrive at a depth of 630 feet; for to that depth the ground is frozen.[34]

Erman's readers felt that it defied common sense to claim that the earth under forests and fields was permanently frozen, so the Russian Academy of Sciences sent a senior member to investigate. After an extensive study, Alexander Theodor von Middendorff concluded that, if Erman was wrong, it was only because he was a careless German, who had underestimated the depth of permafrost in the region, which Middendorff then precisely calculated to be about 754 feet (230 meters). The meticulous quality of Middendorff's research sparked a flurry of studies concerning the phenomenon of frozen Arctic ground, and this new science enabled Russian engineers to begin building the Trans-Siberian Railway in 1891—long before the United States or Canada turned any attention toward northern development.

Today, scientists are quick to point out that as climate change causes temperatures to soar and polar ice to melt, the loss of reflective sea ice increases the temperature of the much more absorbent dark ocean surface surrounding the shrinking ice. This, in turn, accelerates the rate of ice-melt and increases Arctic air temperatures. Such "polar acceleration" causes Arctic temperatures to rise more quickly than they do in southern latitudes. During the weeks before the Baffin Island heat wave, giant ice sheets broke off a nearby ice shelf indicating the process of superheating the Arctic is well underway. But during the previous summer, an unobstructed navigation of the Northwest Passage had already become possible due to the disappearance of so much sea ice. Arctic journalist Ed Struzik describes that season—the summer of 2007—as a turning point for those who study the Arctic. It was then that climatologists began to speculate that distant predictions of an ice-free summer might be radically wrong:

> *Stunned* was the adjective many scientists used to describe how they felt when they saw the satellite data that suggested the Arctic might be seasonally ice-free in as little as five to eight years instead of the 30 to 50 years…predicted only a few years ago. *Shocked* was a word frequently used by others who watched as the old multi-year ice…virtually disappeared.[35]

Without drawing too fine a point on it, it seems safe to say that because we regularly deny or underestimate the reality of climate change's existence, climate change will, by definition, come sooner than we expect. Unfortunately, if we continue to deny the worst the future holds, the consequences may be fatal for our children, unless we find an answer to the question *"Where is the safest place to go?"*

Where to Go

Like one of those terrible dreams of childhood in which you're chased by a nameless danger that snaps at your heels as you continuously never-quite-escape, climate change will grow persistently worse and more comprehensive as the century progresses. Independent citizens determined to survive will likely make several successive moves northward. The safest places will be significant communities in the north that are not isolated, that have abundant water, that have the possibility of agricultural self-sufficiency, that have little immediate risk of forest fires, that are well elevated, and that are built on solid rock. This narrows the field considerably.

The obvious choices are the larger towns of Dawson, White-horse and Yellowknife because they are accessible, and because western portions of northern Canada will experience less severe temperature rise during the coming century. Most importantly, however, precipitation in the Mackenzie and Yukon River basins will increase by 30 or 40% in the coming years, and winter will remain sufficiently cold to kill off the mountain pine beetles (MPB) annually (for a few decades, at least). With money and a determined, coordinated effort, this might be enough time to learn how to eradicate MPBs and similar tree-killing species, like the spruce beetle that threatens Canada's eastern boreal forest.

For their first stop in what will be an ongoing journey, climate migrants should seek temporary safety in one of the Canadian cities strung along the American border like medallions on a concho belt. Beyond the belt, Edmonton is positioned in the middle of Alberta and affords good northern access. These days, there is even a Grey-hound Bus that connects to Yellowknife through the town of Hay River, Northwest Territories. Yellowknife is a good final destination; it alone meets all the criteria mentioned above. Still, my pick for the first step in a staged northward migration would be Vancouver because of its temperate climate and eclectic mix of cultures. It's not a hard place to come to, and who says that escaping climate change has to involve immediate suffering or deprivation? Those who act early may still enjoy some comfort and still have time for further preparations.

Vancouver *will* face its own climate change problems in the coming years, and these will include vulnerability to flooding and impurities in its water supply due to the turbidity created by the extreme storms of winter. But, overall, Vancouver will be a safe and easy first-escape at the end of US I-5; it has the advantages of large social networks of support as well as access to fine northern roads.

Long-range predictions for British Columbia describe a precipitation hole in British Columbia's navel (east of Terrace, northwest of Kamloops and southwest of Prince George). This area will become hotter and drier as lodgepole pine forests die from MPB infestations. There will likely be fires of previously unseen size in this region by 2020 (or roughly around the time of Canada's sesqui-centennial in 2017). Before then, residents of southern B.C. should consider relocating northward.

I favor Prince George for this second stage of the journey be-cause it is a fairly large and well-established community that ac-cesses Canada's only two year-round highways to the north, the Alaska Highway 97/29 and the lesser known Stewart-Cassiar Highway 37. Given my choice, I would take the Cassiar every time since it crosses the breathtaking Stikine (pronounced STI-keen) River, and accesses a dense region of national parks that are flooded with wildlife; it is often compared (foolishly, I think) to Africa's Serengeti Plain. John Muir, my favorite naturalist, described the region in 1879. Forgive me for letting him run on here. He writes so clearly that much of the man still reaches us across the chasm of the intervening years:

> From Glenora to Cassiar, the grasses grow luxuriantly in openings in the woods…on dry hillsides…and over all the broad prairie above the timber-line. A kind of bunch-grass in particular is often four to five feet high, and close enough to be mowed for hay. I never, anywhere, saw finer or more bountiful open pasture. Here the caribou feed and grow fat, braving the intense winter cold, often 40 to 60 degrees be-low zero. What may fairly be called summer lasts only two to three months, winter nine or ten…[and] of pure well-defined spring and autumn there is scarcely a trace. Were it not for the long, severe winters this would be a capital stock county equalling Texas.… I saw thousands of square miles of this prairie-like region drained by tributaries of the Stikeen, Taku, Yukon and Mackenzie rivers.[36]

The beauty and suitability of this spot is enough to tempt any-one, but of course, I am not writing a travelogue. Those journeying north may actually want to avoid the Cassiar Highway because of the road's remoteness since, when mishaps occur, your chances are better if there is traffic. Still, people moving north to escape climate change should not count too much on local help. Resentment will precede and accompany climate migrants, and north of Canada's provinces there is a strong bias against "Southerners" and most Southern things, whether or not they are Yankee in origin.

Even so, in Canada's vast North, only 130,000 people now occupy an area larger than Europe. Northerners are the first to admit that the North needs more people, but there is a likelihood that in the desperate, final stages of a continental migration, established northern residents will find themselves outnumbered and deprived of their innate right to self-determination. The imposition of outside agendas by giant, greedy corporations or by ignorant Southern governments is a dominant theme in the history of Canada's North.

Assisted Migration

For all of the reasons just stated, the best course—for all involved—would be a structured and deliberate colonization that would prepare the ground for immigration, provide transportation, direct colonists to preferred areas, and then assist them once they get there. Although nothing ever came of it, the *Royal Society of Canada* actually made a similar suggestion in 1961:

> It is not irrational to think of colonizing the North. Other countries, mainly the USSR are ahead of us in actual utilization of their boreal territories.[37]

But the former Soviet Union isn't the only country that's encouraged the establishment of migrant colonies. Following World War II, from 1946 to 1967, Finland established 61 new farming communities in the southern portion of Lapland Province in order

> to provide homes for its displaced citizens, to increase economic opportunities for its people, to produce more and...to create a more densely settled zone next to the Soviet border as well as garner votes for...politicians. It did all of these.[38]

Finland's non-coercive model of resettlement in the North is considered one of the world's best examples of a state's expansion into uninhabited areas. Because Finland relied entirely on volunteers and established small farming communities that integrated very well into the network of already existing towns, it avoided many of the social problems—violence, crime and alcoholism—that attended the forced development of large Siberian cities like Ulan-Ude in the former USSR.[39]

But there is also a North American model. In 1934, Franklin Roosevelt created the *Resettlement Administration* to provide opportunities for poor Americans afflicted by the Depression. Over 100 communities were created in the continental United States to relocate out-of-work Americans and provide them with housing, community and jobs. Among these, the relocation of Americans from the upper Midwest to Matanuska, Alaska was a disaster. Although the program also had famous successes like the suburbs, or "Greenbelt Communities," located in Greenbelt, Maryland; Greenhills, Ohio; and Greendale, Wisconsin. The construction of these communities employed 25,000 Americans full-time between the years 1935 and 1938.

But more famous than Roosevelt's successful suburbs, was the colony he created in Alaska that captured America's imagination in 1934 and 1935; it provided a symbol of hope for a country already eager to accept FDR's New Deal. The idea originated with Rex Tugwell, who became head of the Resettlement Administration, and later, governor of Puerto Rico. Tugwell thought to establish a farming community in the fertile rain belt 40 miles northeast of Anchorage. To this end, Roosevelt theatrically halted all homesteading in the area by an executive order on February 4th, 1934. An army of social workers then began to recruit welfare recipients across three extremely depressed northern states: Wisconsin, Minnesota and Michigan. The northernmost area of these states was then known as the "overcut" because the indigenous hardwood forest had been completely harvested, leaving a landscape of dead tree stumps as far as the eye could see. The soil was unsuitable for farming, and there was no other local industry besides lumber.

Alaska's business leaders and the managers of the Alaska Railroad loved the idea of the project and the publicity it attracted. They had been trying to lure settlers for years, knowing that more settlers would be good for business. They hoped the image of a northern agricultural paradise would also end the view many Southerners had that Alaska was a remote and forbidding frontier. They also hoped to jumpstart a badly needed local food industry. Until World War II, Alaska was completely reliant on Southern goods, which meant that a lot of Alaskan money poured southward, out of the state. But this was 1935, eight long years before the Alaskan Highway was finally completed. Alaska's remoteness presented a formidable obstacle to the Alaska Relief and Rehabilitation Corporation that handled logistics for the Matanuska settlers.

Successful farmers from Minnesota, Wisconsin and Michigan were excluded from participating in the project. Since the idea was to create jobs for out-of-work Americans, the relief workers selected applicants only from among the welfare recipients who applied. Very few applicants had any real farming experience. There have been claims that the relief workers selected the worst welfare cases among the applicants in order to be rid of them. But this opinion seems to be an expression of native Alaskans' resentment of the migrants, who enjoyed a very open-hand on the part of the federal government. "Coming into the country" for the Matanuskans was substantially easier than what most ordinary Alaskans had endured, since many were children born to the migrants who stayed in the North following the invasion of Alaska and the Yukon during the Klondike Gold Rush. To be fair, the native-born Alaskans' resentment was not completely ill-founded. The Resettlement Administration was so generous that Tugwell resigned in 1936 under criticism he had created a socialist commune in the United States. Maybe he had. Once selected, the fortunate Matanuska settlers were immediately given a loan of $3,000 and provided with 40- to 80-acre plots, for which they had

> to make payments over a thirty-year period at an annual interest rate of 3%. The federal government agreed to build...
> houses and barns and to pay for transporting families and

up to 2,000 pounds of their household goods.... Farm machinery, equipment, livestock and supplies were made available by the corporation to the colonists for purchase, lease
or payment for use. Supplies could be obtained as needed at
cost until the colonists became self-sufficient. Educational,
cultural, recreational, and health services were [also]...pro-
vided.[40]

Although it was not widely publicized in the 1930s, about five million Depression-era dollars were spent relocating 201 families to
Alaska. Unfortunately, apart from the rail and ship journey to the
North (which was something of a national event receiving almost
daily coverage in newspapers, on the radio, and in newsreels) the
colonizing effort itself was badly planned and poorly executed.
When the families arrived in the Matanuska Valley on May 10,
1935, their houses had not yet been built, and the rainy season had
already begun. The Southerners lived in tents surrounded by a sea
of mud for months, having missed the opportunity to plant anything that year. Nine families left in disgust by July; more soon
followed. By 1948, only 31% of the original settlers and 43% of their
replacements still lived and farmed in the Matanuska Valley.[41]

Still, some of the colonists understood very clearly that they
had little in the South to return to and that what they were being
given was a genuine opportunity to create a new life and a new
livelihood. These hardy, practical types endured the early years of
bureaucratic chaos and eventually managed to develop a local Alaskan food industry (albeit one that received a boost by providing
meat and vegetables to the US Army garrisoned in Anchorage after
1942). The point, however, is not the degree of success or failure the
Matanuskans enjoyed, but the usefulness of their example. Governments looking to organize a northern migration and plan a colony
could learn a lot from the plain-dumb, first-draft mistakes made by
the New Dealers in Alaska in the 1930s.

Matanuska is especially valuable since there are few things with
more instructive force than that of a truly bad example. But there

is more help from another source. These days, conservationists talk a lot about "assisted migration." Usually it means helping a local animal species cross a man-made boundary (like a road or a city) to reach a more suitable environment. The *bay checkerspot butterfly* is most often used as an example. Once commonplace around the San Francisco area, climate change, invasive plants and urban development have all but destroyed the grasslands in which this two-inch orange, black, and white beauty breeds. And we know that, as the temperatures of northern California rise, there will be shifts in rainfall that will devastate the checkerspot population by depriving them of the dwarf plantains and owl's clover that are the caterpillars' host plants during the vital growth period before they become chrysalides.[42] Pointing to the successful reintroduction of grey wolves into Idaho, Wyoming and Montana, some biologists now recommend "assisting" the checkerspots by transplanting caterpillars to a cooler climate, like that of Washington state. The controversy has acquired some urgency because many biologists believe that between 15 and 37% of all species now living will become extinct by 2050. But simply moving a single species from one spot to another is no guarantee of success. Conservation biologists would have to move a whole network of interrelated species in order to guarantee familiar food resources like the plantains and clovers on which checkerspots thrive.[43]

In this sense, assisted migration for human beings is a much more practical solution than it is for flora or fauna. Human beings possess "culture," which includes a vast body of skills, technology, history, stories and attitudes enabling us to adapt readily in new environments. The Finns knew this and cleverly assisted migrants to each of their 61 new settlements by establishing the new communities close to existing ones so experienced residents could provide expertise and assistance to the newcomers by sharing their coping strategies and culture of survival. Essentially, similar tactics enabling each new generation to benefit from long, local experience underlies the tradition of a "council of elders" among all indigenous people, including those of the North.

Access by road has made construction of new colonies less time-consuming, challenging and costly than during any previous period. This access also reduces the cost of Southern goods to communities serviced by these year-round roads. So the lessons of Matanuska, of Finland, and of conservation biology could be very instructive to contemporary Canadian stakeholders making preparations to accommodate migrants fleeing climate change on our continent in the current century. Such preparations are already being made by the world's largest countries, which will share humanity's future with Canada and the United States at the top of the world. The plans for northern development and migration by Asia's giants are largely secret. But Russia is clearly laying claims to minerals on the Artic seabed while preparing a fleet of atomic-powered icebreakers to dominate the Arctic shipping lanes. Simultaneously, China has been sending exploratory naval missions to the unpopulated islands of Canada's Franklin District, sensing the promise of new fishing grounds and new and empty lands.[44]

The new North will need a substantial and productive population to maintain its independence from Russia in the coming century, as well as to ensure that Russia, the most populous Arctic nation and the one that already controls most Arctic natural gas resources, does not dominate a lion's share of the pan-polar region. The best way to achieve these ends will be to create a northern ark (or system of arks) for environmental refugees fleeing the reduction of human habitat in the lower 48 states and in southern Canada. This is something concrete that we can do now to fight the worst danger of climate change: massive numbers of human deaths during the coming period of social and agricultural collapse and continental migration.

The Herculean mobilization required to shape the new North will require everyone's effort and every resource we have, but nothing less than our children's survival and the survival of our species is at stake.

Cooler Climes
and Higher Ground

*The argument from global warming to [economic] growth
reduction is so weak that one reaches for deeper explanations
of its persistent appeal.... Most climate radicals are also
passionate haters of greed and luxury, people who in previous
ages might have been Cromwells or Savonarolas. Infusing
a good deal of environmentalist literature is a love of the hair
shirt. The puritan accent is unmistakable in George Monbiot's
announcement that his is a campaign "not for abundance but
for austerity...a campaign not for more freedom, but for less...
a campaign not just against other people, but also against
ourselves." Here...lies the beating heart of climate activism.*

LORD RICHARD SKIDELSKY, EDWARD SKIDELSKY,
HOW MUCH IS ENOUGH? (2012)[1]

2012 is a bit late in the day for entrenched energy-capitalists to challenge anything intended to mitigate climate change. Still, the family
fortune of the English Lord quoted above (a man famous for his
eccentric opinions) began with his parents' pre-war acquisition of
a coal mine in China. Clearly, his concern about the price of mitigating global warming begs the question of defensive self-interest:

Its [anthropogenic climate change's] elimination requires the
abandonment not just of this or that luxury but of coal, gas

and oil—the lifeblood of industrial civilization. This provides a convenient platform to those who never liked industrial civilization…[like] George Monbiot.[2]

Likely, the Skidelsky co-authors are reluctant to acknowledge contributions by their parents and grandparents to atmospheric CO_2 and to the climate crisis of the modern world. Quite correctly, they observe: "Any radical shift in consciousness requires the stimulus of crisis."[3] Unfortunately, they miss what much of the rest of the world has already perceived. The carbon summer is upon us, growing worse and more intense with each passing year through droughts, forest fires, heat waves, sea level rise and extreme storms. Even in the United States, where climate change used to excite a completely negative response from conservatives, the tide is turning. A recent Pew poll found 44% of registered Republicans believe in climate change, while a Gallup poll found that 40% of Republicans actively worry about climate change. In bi-partisan terms, nearly 70% of Americans believe that climate change is real and is happening today.[4]

The figures are slightly higher in other polls. A recent Yale/George Mason University survey, for example, found that 52% of Republicans now believe in climate change.[5] Researchers at the University of Texas found that 73% of respondents (Texans all) agree climate change is very real.[6] Whatever discrepancies there are among the various figures from various sources, it is now safe to say that much like evolution, global warming is now widely—if somewhat reluctantly—accepted.

Two further beliefs have become commonplace. These are that climate change is "man-made" (or anthropogenic) and that contemporary climate changes around the world are happening much faster than we predicted or hoped. Although these beliefs were not common at the time, I began writing this book in 2009 convinced of all three points: climate change is real; it is man-made; it is happening faster than we thought. From these presuppositions, *American Exodus* puts forward two connected points. The first of these is that migration is a time-honored response to climate change. It has happened many times in the past. Consequently, even here in North America, climate migration is a consistent historical theme with many climate migration events occurring in the 20th and early 21st centuries.

My second argument is more speculative, and if that puts me in the company of "climate radicals" who write as persuasively and as well as George Monbiot, so be it. *American Exodus*'s second claim is simply that since an era of extreme climate change resulting from anthropogenic global warming has now begun *definitively*, we can expect much more human and animal migration to follow. So, for example, there are robins well above the old treeline along the Arctic coastline in Tuktoyaktuk. Local journalists find themselves explaining what these strange red birds are to the local residents.[7] This, of course, is charming, but as America's Southwest dehydrates and its northeastern shorelines erode, I anticipate many more *human* migrants will seek out cooler climes and higher ground. Canada, of course, is the obvious destination for Americans suffering from the increasingly "hot, flat and crowded" conditions of the United States in the 21st century.

At this point you might say "Hey, wait a minute. Is there any evidence today that climate migrations will happen?"

Fair question.

The stark-naked, unvarnished answer is "yes." In Alaska, sea level rise and storm surges resulting from the disappearance of the coastal icepack are already forcing the residents of coastal communities to remove inland. The Army Corps of Engineers is now moving the homes of 350 residents of Newtok away from their vulnerability to storm surges and erosion. Suzanne Goldenberg writes:

> There is no disputing the real-time effects of climate change. Alaska is warming faster than anywhere else in America, setting off a circumpolar scramble for oil and other resources given up by the melting ice and threatening the livelihood of those who still live off the land and the sea.[8]

Although Goldenberg is exclusively describing the plight of New-tok, there are many other Alaskan communities affected in similar ways including Dillingham, Kaktovik, Kivalina, Koyukuk, Shak-toolik, Shishmaref, and Unalakleet. In the Canadian Arctic, the coastal towns of Tuktoyaktuk, (aka "Tuk") NWT, and Pangnir-tung, on Baffin Island in the new Nunavut territory, are experienc-ing the same problem, as are Yukagir communities on the coast of Siberia. Further relocations in Alaska and throughout the Arctic are now just a matter of timing and budgets.[9]

Meanwhile, Goldenberg calls the unlucky Alaskans who will be relocated to higher ground "America's First Climate Refugees." This is my only quibble with Ms. Goldenberg's outstanding en-vironmental journalism, and it is a very small one. The people of Newtok are not the first climate refugees—either in North America or in the United States of America. Climate change has occurred many times during the history of our continent. There is a strong possibility that climate change forced or enabled the migration of Asian peoples across a now submerged land-bridge into what is now Alaska 10–15,000 years ago. Since then, climate change has af-fected many American peoples, including the Maya of the Yucatan and the Anasazi of the Four Corners.

Although they fascinate me, I do not write about these ancient groups. I am the father of three boys, and I am concerned about survivability in the frightening new world they will inherit. *Ameri-can Exodus* begins in the 20th century with modern Americans in the 1930s. These people suffered through the environmental catastrophe we now call the Dust Bowl. Between the years 1929 and 1940, 2.5 million people left the Great Plains as drought and dust storms made this semi-arid region unsuitable for farming of any kind. It makes more sense to call *these* people "America's First Climate Refugees." Because their struggle remains within living memory, the lessons of their migration are still relevant, relatable, and real.

Outmigration from the Great Plains continues today, but now the culprit is lack of water. The Ogallala fossil aquifer has been so drawn down over the years that farming and human habitability have been radically reduced throughout the region. Climate change has made La Niña a nearly permanent condition that brings with it nearly permanent drought. So many people have now left the Plains that in many areas it has returned to unpopulated wilderness.

Geographically, Mexico is part of North America, and, as the Great Plains are being abandoned, northern Mexico is simultaneously being desiccated by the poleward expansion of the tropics, a direct result of global warming. This began in a noticeable way at a very bad economic time. In 1982, the national economy of Mexico collapsed, making subsistence farming very difficult. Desperately, Mexicans cut down their forests, drew down their aquifers, farmed unsuitable land, and abandoned the tradition of "three sisters" crop rotation that had made life possible on small Mexican farms. In addition, they sold their land in order to pay down debt or to support themselves. At the same time, climate change and global warming were dehydrating the North. In response to worsening environmental and economic conditions, many rural Mexicans became urban "chilangos," fleeing to jobs in Mexico City.

Then, as "la crisis" deepened in the 1990s, and as jobs and life became too hard even in the capital, more and more Mexicans fled north to the United States, where between 11 and 20 million Mexican-born immigrants (legal and illegal) now live. Economic historian Sing C. Chew has demonstrated that climate change is always accompanied by economic collapse; for this reason, the Mexicans who live and work in the United States can also be thought of as climate refugees or climate migrants. Their unwelcome reception in the United States is a valuable lesson to future climate migrants, as is the increasing impact of Latino migrants on American political and cultural life.

The Mexican migration slowed a bit in the first decade of the 21st century. But in 2005, 500,000 American citizens became climate migrants after New Orleans was swamped by Hurricane Katrina. The tepid response of FEMA to the plight of un-evacuated New Orleanians was a repetition of the lifeboat ethics that prevailed in

California during the Dust Bowl, when climate refugees fled the Great Plains and sought employment in the Golden State. The characteristic knee-jerk habit of blaming disaster victims for their own plight has impeded the reconstruction of New Orleans for eight long years. In coming years, we can expect more people to flee flooded coastal cities.

Moreover, across the Midwest and northern states, heat waves increased in frequency, intensity and duration during the final years of the 20th century. In 1995, a heat wave lasting five days caused at least 800 deaths in Chicago, with even more deaths logged in the nearby cities of St. Louis and Milwaukee. As global warming increases, the intensification of heat waves becomes a spectacularly lethal phenomenon throughout the Northern Hemisphere as, for instance, when the European Heat Wave of 2003 caused between 30,000 and 50,000 deaths in 2003.

Since then, there have been intense European heat waves in 2006 and 2010. The United States suffered killing heat waves in 2010, 2011 and 2012. With urban infrastructure now crumbling from lack of investment, the increasing duration of heat waves represents a serious challenge to the habitability of major urban centers in the Northeast. At such times, the burden which air conditioning places on the electrical power grid leads to inevitable power outages. In turn, lack of refrigeration interrupts the food and water supplies, as it did in Chicago in 1995. Without air conditioning, many isolated seniors and children die from hyperthermia, which is by far the largest annual environmental killer. Simply put, in the future the largest cities of the Northeast and Europe will become unlivable for several months each year. Writing in 2012, only Lord Skidelsky disagrees:

> It is not self-evident to dispassionate eyes, that global warming requires us to abandon growth. It might rather require us to persevere with growth, so as to finance the technologies needed to mitigate its consequences.[10]

What utter nonsense! If the time for developing new technologies and for mitigating global warming has not yet run out, it is running out quickly. We have had 40 years of warnings, and we have not built any arks. There is only a bit of time left to prepare. The economy of the South and Southwest has been in decline since 2005, when a killing drought began impeding cotton and tobacco production and reducing the number of beef cattle and sheep. The Mississippi River is so low it can no longer support barge traffic. Finally, in 2011, a group of climate scientists were able to connect that year's drought to climate change using statistical probability. In their words:

> One analogy of the effects of climate change on extreme weather is with a baseball player…who starts taking steroids and afterwards hits on average 20% more home runs in a season than he did before…. You would be able to [say]… that all other things being equal, steroid use had increased the probability of [home run hits] by 20%.[11]

The home run in this case involves the eradication of most of Texas's forests, including the death of 500,000 trees from lack of water and the lethal megafires that followed. Still, the utility of the statistical method of determining connections between climate change and extreme weather has illuminated the risks insurance companies face through the increasing frequency and intensity of extreme weather. For insurance companies, global warming and climate change are cold, hard, mathematical facts as unforgiving as red ink on an annual budget. Since 1980, Munich Re notes that yearly losses from climate-related disasters have skyrocketed from a few billion to more than $200 billion (in 2010 US dollars). The United States now averages $68 billion in losses each year due to extreme climate change weather events. Estimates suggest $80 billion as the total price tag for extreme weather events in 2014.

In May 2013, a rash of 28 deadly twisters spread across Tornado Alley in the Southwest. Tornadoes have increased in frequency and intensity with global warming. The damage they cause has increased, also. In Joplin, Missouri, in 2011, a tornado caused a record $2.5 billion in damage. Earlier this month (May 4, 2013), a tornado in Granbury, Texas, narrowly missed being designated EF5, the most powerful and destructive tornado designation. The Granbury twister was rated EF4 since its winds reached, but did not exceed, 200 miles per hour. Extreme EF5 storms can now reach wind speeds of 312 mph. I am writing these words on May 22, 2013. Until yesterday, there were only eight known tornadoes with the highest (EF5) rating. Four EF5 tornadoes occurred in the late 19th and 20th centuries. The other half of the EF5 killer-tornadoes (including the Joplin storm) all occurred in the late winter and spring of 2011. Yesterday, however, a new EF5 struck Moore, Oklahoma, a suburb of Oklahoma City, and it was as though a baseball player on steroids once again hit it out of the park. Consequently, the majority of the most powerful tornadoes have now occurred in the second decade of the 21st century. Clearly, accelerated global warming is happening today. Clearly also, the problem needs to be recognized, acknowledged, publicized and addressed with the same urgency we would devote to an invasion.

In the fall of 2012, Superstorm Sandy confirmed that once-in-a-century northern Atlantic hurricanes (and storms) will increase in frequency and intensity. As a result, storm surges will cause greater and greater damage along the densely populated shores of America's East Coast. This issue is of vital concern in the United States, where the hotspot along the Atlantic coast puts at risk nearly half of continental America's entire population. Storm surges are equally important in Mexico and Canada. In Tobasco, Mexico, 1,000,000 people were relocated from their homes during the storm surge flooding of November 2007. In Canada, the only country surrounded by three oceans, the problem is even more serious since over 240,000 kilometers of Canadian coastline (the largest of any nation) leaves many communities vulnerable, including those along the shore of the Arctic Ocean. On the Atlantic coast, the town of Gaspé (on the Gaspé Peninsula) is repeatedly flooded by the storm surges that overflow Rivière-au-Renard. Like Newtok, the entire town will have to be relocated.

Depending on which source you believe, Superstorm Sandy cost the United States a whopping $50–80 billion in damages. (The high-end estimates include figures related to business losses created in the aftermath of the storm.) On average, America now pays out $68 billion per year (since 2011) to repair the infrastructure damaged by extreme weather events like Superstorm Sandy. Insurers (like Munich Re) are quick to connect climate change to global warming, and they are quick to observe that these costs will only increase in coming years.

Unfortunately, the monetized damages of climate change's extreme weather events do not tell the real story. In coming years, many human lives will be lost as the secure livelihoods characteristic of the *Pax Americana* era come to an abrupt and painful end. The collapse of the three North American economies, coupled with the continuous and increasing damage to the infrastructure of our societies, will together make our large, complex countries unaffordable. If we take an honest view, we can see that collapse is happening right now in Mexico. "The implication is clear," writes Joseph Tainter, a Cambridge archeologist who studies the collapse of complex societies, "civilizations are fragile, impermanent things."[12] Social collapse has happened many times before. It is as inevitable as seasonal change. Still, if we prepare for it, the end our complex national structures does not need to mean the end of humanity.

Charles Darwin described his view of the end of humanity many years ago, writing:

> Believing as I do that man in the distant future will be a far more perfect creature than he now is, it is an intolerable thought that he and all other sentient beings are doomed to complete annihilation after such long-continued slow progress. To those who fully admit the immortality of the human soul, the destruction of our world will not appear so dreadful.[13]

I am a Unitarian. I am not convinced of the immortality of the human soul, and I believe in the vital importance of our continued existence. Speaking self-centeredly, as though I were a perishable leaf who believes in the significance and metaphysical value of his own tree—the human tree, the tree-of-life—it would be dreadful to come to the end of mankind. It would spread meaninglessness, futility and silence throughout the universe. If there is a pure, re-fined evil common to all mankind, surely this is it. Let's get to work now to prevent that horror. In the meantime too, let's inspire our children to have faith in their own power and competence so that they will be able to meet the challenges and continue the fight.

It will last for generations.

*If there were no such thing as climate-induced migration
(CIM), then there would be nothing to contemplate.
Yet climate change has occurred—and is likely a growing
contributor to trans-border movements—so international
society's response becomes a matter of deep concern....
If climate change is forcing increasing numbers of people
to cross international borders in search of safety and refuge....
CIM is an ethical concern of the first order.*

GREGORY WHITE, CLIMATE CHANGE AND MIGRATION:
SECURITY AND BORDERS IN A WARMING WORLD.

Notes

Introduction

1. Frank W. Turner (New York: Penguin, 1996). Originally published as *Geronimo's Story of His Life* in 1906. Read this book!
2. Instituto Nacional de Estadística, Geografía e informática (INEGI), "Estatísticas a Propósito del Día Mundial del Agua," Mar 22, 2006, inegi.gob.mx/, accessed Feb 15, 2013.
3. Elizabeth Deheza, Jorge Mora, "Climate Change, Migration and Security: Best-Practice Policy and Operational Options for Mexico," *Royal United Services Institute for Defence and Security Studies* Whitehall Report Series, Jan 31, 2013.
4. Mark Carrasco et al., "Quantifying the Extent of North American Mammal Extinction Relative to the Pre-Anthropogenic Baseline," *PloS One* Vol 4 no. 12, Dec 16, 2009; Richard Seager et al., "Model Projections on an Imminent Transition to a More Arid Climate in Southwestern North America," *Science Magazine* Vol 316 no. 5828, 2007: 1181.
5. R. Bronen, "Climate-Induced Relocations: Creating an Adaptive Governance Framework Based in Human Rights Doctrine," *New York University Review of Law and Social Change* 35 (2011): 372.

Chapter 1

1. "Dust Storm Shuts Down Interstate in Oklahoma," Oct 19, 2012, usatoday.com, accessed Oct 19, 2012.
2. David A. Hodell et al., "Possible Role of Climate in the Collapse of Classic Maya Civilization," *Nature* (letters) Vol 375, June 1, 1995: 391–94.
3. Donald Worster, *Dust Bowl: The Southern Plains in the 1930s.* (New York: Oxford University Press, 1979): 24.

4. Aldo Leopold, "The Conservation Ethic," *Journal of Forestry* Vol 31, Oct 1933: 634.

5. James C. Malin, "Dust Storms, 1850–1900," Part 2, *Kansas Historical Quarterly* Vol 14, Aug, 1946: 267.

6. Charles M. Hudson, *The Southeastern Indians*. (Knoxville: University of Tennessee Press, 1976): 127, 122–83; but see also his *Elements of Southeastern Indian Religion*. (Boston: Brill Academic, 1984).

7. Mathew Paul Bonnifield, *The Dust Bowl: Men, Dirt and Depression*. (Albuquerque: University of New Mexico Press, 1979): 42.

8. Worster, (1979): 6.

9. Paul Sears, *Deserts on the March*. (Norman: University of Oklahoma Press, 1959): 67.

10. H. H. Bennett, "Land Abuse," *New York Times* Mar 19, 1936: E7.

11. Darden Asbury Pyron, *Southern Daughter: The Life of Margaret Mitchell*. (New York: Oxford University Press, 1991): 56.

12. Bonnifield, *The Dust Bowl*: 51.

13. Craig Canine, *Dream Reaper: The Story of an Old-Fashioned Inventor in the High-Tech, High-Stakes World of Modern Agriculture*. (New York: Knopf, 1995): 154.

14. Bonnifield, *The Dust Bowl*: 49.

15. Donald Worster, *The Wealth of Nature: Environmental History and the Ecological Imagination* (Oxford University Press, 1993): 64; Worster, *Dust Bowl* (1979): 13.

16. Worster (1979): 64.

17. Bonnifield, *The Dust Bowl*: 49–50.

18. Giles Slade, *Made to Break: Technology and Obsolescence in America*. (Cambridge: Harvard University Press, 2006): 66–67.

19. James N. Gregory, *American Exodus: The Dust Bowl Migration and Okie Culture in California*. (New York: Oxford University Press, 1989): 6–7.

20. Worster (1979): 42.

21. Walter J. Stein, *California and the Dust Bowl Migration*. (Westport: Greenwood Press, 1973): 32–33.

22. Paul J. Taylor, "Drought Refugee and Labor Migration to California, June–December 1935," in Paul J. Taylor, *Labor on the Land, Collected Writings 1930–1970*. (New York: Arno Press, 1981): 125.

23. Gregory, *American Exodus*: 10.

24. "Poverty Drives Out Drought Refugees," *Washington Post* Jan 17, 1931: 3.

25. Worster (1979): 17.

26. Ibid.

27. Article 1A, 1951 UN Convention Relating to the Status of Refugees.

28. Fabrice Renaud et al., "Control, Adapt or Flee: How to Face Environmental Migration," United Nations University, *Intersections* no. 5 (2007): 10. Accessed at ehs.unu.edu, Oct 5, 2007.

29. Ibid.

30. Sing C. Chew, *World Ecological Degradation*. (Walnut Creek, CA: AltaMira Press, 2001): 57.

31. Myers (1997): 167–68.

32. Caroline A. Henderson. Letter dated June 30, 1935, *The Atlantic Monthly* Sep 1936: 543.

33. A. J. McMichael et al., "Climate Change and Human Health: Present and Future Risks," *The Lancet* 367 (9513) 2006: 859–69.

34. Gregory, *American Exodus*: 7.

35. Worster (1979): 6.

36. Stein, *California and the Dust Bowl Migration*: 37.

37. Carey McWilliams, "Getting Rid of the Mexican," *American Mercury* (28) Mar 1933: 322–24.

38. "Lack of Jobs is Driving Thousands of Mexican Workers across the Border," *Christian Science Monitor* May 4, 1931: 1.

39. Stein, 73.

40. Leonard J. Leader, *Los Angeles and the Great Depression*. (New York: Garland Publishing, 1991): 202. This strange document is a reprint of the 1971 doctoral thesis by a practicing journalist who decided to complete a PhD at UC Berkeley when he was 52. No doubt as a result of Dr. Leader's early professional training, it is a much more interesting, informative read than any other "dissertation book" I've encountered.

41. "Immigrant Curb Splits California," *New York Times* Feb 9, 1936: E11.

42. "Los Angeles Police Continue to Stem Immigrant Horde," *Christian Science Monitor* Feb 5, 1936: 2.

43. Stein, 74.

44. Leader, *Los Angeles and the Great Depression*: 202.

45. Stein, 75–77; Gregory, *American Exodus*: 88.

46. Stein, 57.

47. Cited in Gregory, *American Exodus*: 95.

48. Stein, 63.
49. Stein, 60.
50. Stein, 63.
51. *Los Angeles Daily News* Mar 3, 1939, cited in Gregory, 98.
52. Garrett Hardin, "Lifeboat Ethics: The Case against Helping the Poor," *Psychology Today* Sep 1974, retrieved online, Nov 6, 2007 at garretthardinsociety.org; Garrett Hardin, "Living on a Lifeboat," *BioScience* Vol 24 (10) 1974: 561–56, retrieved online Nov 6, 2007 at garretthardinsociety.org.
53. Eliphaz in Job 5:6, but see also 4:17 (Eliphaz), "shall mortal man be more just than God? Shall a man be more pure than his maker"; and 8:6 (Bildad) "if thou wert pure and upright surely he would awake for thee, and make the habitation of thy righteousness prosperous"; (and, to a lesser degree, 11:20 [Zophar] "the eyes of the wicked shall fail, and they shall not escape, and their hope shall be as the giving up of the ghost").
54. Michael Lerner, *The Belief in a Just World: A Fundamental Delusion.* (New York: Plenum Press, 1980): 105.
55. John Steinbeck, *John Steinbeck. The Grapes of Wrath and Other Writings.* (New York: Library of America, 1996): 510.
56. Extemporaneous speech July 13, 1940, Culbert Olson Papers, Bancroft Library, University of California, Berkeley, Carton 6; cited in Stein, 112.
57. Kurt M. Campbell et al., eds., *The Age of Consequences: The Foreign Policy and National Security Implications of Global Climate Change.* (Washington, DC: Center for Strategic and International Studies (CSIS); Center for New American Security (CNS), Nov 5, 2007): 39.
58. Ibid.

Chapter 2

1. S. Feng et al., "Linkages among Climate Change, Crop Yields and Mexico-US Cross Border Migration," *Proceedings of the National Academy of Sciences* 107.32, Aug 10, 2010: 14257–62.
2. Ilan Stavans, "Autobiographical Essay," In Ilan Stavans, ed., *Becoming Americans: Four Centuries of Immigrant Writing.* (New York: Library of America, 2009): 630.
3. Jose Emilio Pacheco, "Badlands," in Ilan Stavans ed., *The FSG*

Book of Twentieth Century Latin American Poetry. (New York: FSG, 2011): 531.

4. Gregory B. and John R. Weeks, *Irresistible Forces: Explaining Latin American Migration to the United States.* (Albuquerque: University of New Mexico Press, 2010): manuscript pages 20–21.

5. "Calderón: Mexico Drug Gangs Seek to Replace State," BBC.com Aug 5, 2010; retrieved online on Aug 10, 2010 from bbc.co.uk.

6. Tracy Wilkinson, "Calderon Delivers Blunt View of Drug Cartels' Sway in Mexico," Aug 10, 2010, *Los Angeles Times.* Retrieved online Aug 10, 2010 at latimes.com.

7. Marcela Cerruta and Douglas Massey, "Trends in Mexican Migration to the United States: 1965–1995," in Jorge Durand, Douglas Massey, eds., *Crossing the Border: Research from the Mexican Migration Project.* (New York: Russell Sage, 2004): 21.

8. Judith Adler Hellman, *Mexico in Crisis,* second edition. (New York: Holme and Meir, 1983): 223–24.

9. Ruben Martinez, *Crossing Over: A Mexican Family on the Migrant Trail.* (New York: Henry Holt, 2001): 10.

10. Scott Whiteford and Roberto Melville, eds., *Protecting a Sacred Gift: Water and Social Change in Mexico.* (San Diego: Center for U.S.-Mexican Studies, 2002): 8.

11. Ibid.

12. Cynthia Hewitt de Alcantara, *Modernizing Mexican Agriculture: Socioeconomic Implications of Technological Change 1940–1970.* (Geneva: UN Research Institute for Social Development, 1976): 310.

13. Joel Simon, *Endangered Mexico: An Environment on the Edge.* (San Francisco: Sierra Club Books, 1997):37.

14. Ibid.

15. Hellman, *Mexico in Crisis:* 230.

16. Feng et al., "Linkages": 107. Retrieved online on July 30, 2010 at pnas.org.

17. Ibid.

18. Barry Sinervo et al., "Erosion of Lizard Diversity by Climate Change," *Science* Vol 328, no. 5980, May 14, 2010: 894–99.

19. Judith Adler Hellman, *The World of Mexican Migrants: The Rock and the Hard Place.* (New York: The New Press, 2008): 38.

20. Matthew J. Gibney et al., *Immigration and Asylum: From 1900 to the Present.* (Santa Barbara, CA: ABC-Clio, 2005): 77.

21. Maria Cristina Garcia, *Seeking Refuge: Central American Migration to Mexico, the United States and Canada*. (Berkeley: University of California Press, 2006): 45.

22. "The New Ellis Island," *Time Magazine* June 13, 1983: 18–20.

23. Christian Parenti, *Tropic of Chaos: Climate Change and the New Geography of Violence*. (New York: Nation Books, 2011): 186–87.

24. Cynthia Gorney, "A People Apart: The Tarahumara," *National Geographic* Nov 2008. Retrieved online on Sep 18, 2012 at nationalgeographic.com.

25. Kendrick Blackwood, "Language Barrier Turned Hospital into Prison," Aug 13, 2000. Retrieved online on Sep 18, 2012 at www2 .ljworld.com.

26. Jorge Castañeda, *Ex Mex: From Migrants to Immigrants*. (New York: The New Press, 2007): 125.

27. William Booth, "Mexico At War," *Washington Post* July 4, 2010. Retrieved online on July 4, 2010 at washingtonpost.com.

28. Alejandro Portes, Robert L. Bach, *Latin Journey: Cuban and Mexican Immigrants in the United States*. (Berkeley: University of California Press, 1985): 114.

29. Kathryn Ferguson et al., *Crossing with the Virgin: Stories from the Migrant Trail*. (Tucson: University of Arizona Press, 2010).

30. Joseph Contreras, *In the Shadow of the Giant: The Americanization of Modern Mexico*. (New Brunswick, NJ: Rutgers University Press, 2009): 61.

31. Ibid., 79.

32. Sing C. Chew, *Ecological Futures*. (Lanham: AltaMira Press, 2008): 45. See also Introduction to Sing C. Chew, *Recurring Dark Ages*. (Lanham: AltaMira Press, 2007): xvi.

33. Stephen Mumme, "Mexico's New Environmental Policy: An Assessment," in Donald Schulz et al., eds., *Mexico Faces the 21st Century*. (Westport, CT: Praeger, 1995): 98.

34. Introduction, in Scott Whiteford and Roberto Melville, eds., *Protecting a Sacred Gift: Water and Social Change in Mexico* (San Diego: Center for U.S.-Mexican Studies, 2002): 9.

35. Mumme, "Mexico's New Environmental Policy."

36. Ibid.

37. Associated Press, "Mexico: 28,000 Killed in Drug Violence since 2006." Retrieved online Aug 4, 2010 at google.com/hostednews /ap.

38. Weeks and Weeks, *Irresistible Forces: Explaining Latin American Migration to the United States.* manuscript page 9.

39. CNN.com, "Mexico Files Court Brief against Anti-immigration Law." Retrieved online on June 22, 2010.

40. Amanda Lee Myers, "Immigrant Deaths in Arizona Desert Soaring in July," July 16, 2010. Retrieved online at yahoo.com.

41. Leo R. Chavez, *The Latino Threat: Constructing Immigrants, Citizens and the Nation.* (Stanford, CA: Stanford University Press, 2008): 154.

42. Ibid., 160.

43. Ibid., 164.

44. Ibid., 165.

45. Don Poulson, "The Way I See It: The Not-So-Great Business of California," *Daily News*, retrieved online on redbluffdailynews.com.

46. Richard Simon, "California Could Lose a House Seat after 2010 Census," *Los Angeles Times*, retrieved online Aug 17, 2009 at articles.latimes.com.

47. This is a pretty well documented fact. See for example William Frey's "Metropolitan Magnets for International and Domestic Migrants," in Bruce Katz et al., eds., *Redefining Urban and Suburban America: Evidence from Census 2000*, Vol 1 (Washington, DC: Brookings Institute, 2005): 9, 17; Arthur Laffer et al., eds., *The End of Prosperity: How Higher Taxes Will Doom the Economy—If We Let It Happen.* (New York: Simon and Schuster, 2008): 161; Frank Bean et al., eds., *Immigration and Opportunity: Race Ethnicity and Employment in the United States.* (New York: Russell Sage, 2003): 324.

48. Candice Reed, "Dear California, I'm Dumping You," *Los Angeles Times* Aug 16, 2009. Downloaded on Jan 26, 2013 from articles .latimes.com.

49. Peter Schrag, *California: America's High Stakes Experiment.* (Berkeley: University of California Press, 2006): 109.

50. Norris Hundley, *The Great Thirst: Californians and Water: A History.* (Berkeley: University of California Press, 2001): 44.

Chapter 3

1. Kevin E. Trenbreth et al., "Estimates of the Global Water Budget and Its Annual Cycle Using Observational and Model Data," *Journal of Hydrometeorology* Vol 8 (2007): 758–69; cited in Wallace

Broecker and Robert Kunzig, *Fixing Climate: What Past Climate Changes Reveal about the Current Threat—And How to Counter It.* (New York: Hill and Wang, 2008): 161.

2. Alan Grainger, *The Threatening Desert: Controlling Desertification.* (London: Earthscan, 1990): 35–37.

3. W.E. Smythe, *The Conquest of Arid America.* (Seattle: University of Washington Press, 1969): 31–32.

4. Kevin E. Trenbreth, "A Changing World for Climate Research," *Geotimes* Dec 2005: 24–25; personal e-mail message, June 2008.

5. Elizabeth Royte, *Bottlemania.* (New York: Bloomsbury, 2008): 15.

6. Martin Hoerling, Arun Kumar, "The Perfect Ocean for Drought," *Science* Vol 299, Jan 31, 2003: 691.

7. Broecker and Kunzig, *Fixing Climate*: 179.

8. Richard Seager et al., "Model Projections for an Imminent Transition to a More Arid Climate in Southwestern North America," *Science* 316 (2007): 1181–84.

9. Mark Cane et al., "Twentieth Century Sea Surface Temperature Trends," *Science* 275 (1997): 957–60.

10. Anne Matthews, *Where the Buffalo Roam.* (New York: Grove Press, 1992): 21.

11. Ibid., 21; 23–24, mss 93.

12. Ken Midkiff, *Not a Drop to Drink: America's Water Crisis (And What You Can Do).* (Novato, CA: New World Library, 2007): 28, 30.

13. Monica Barnes, David Fleming, "Filtration Gallery Irrigation in the Spanish New World," *Latin American Antiquity* Vol 2, no. 1 (Mar 1991): 48. The simplicity and success of this ancient technology encouraged its spread across the trade routes into Africa, China, the Middle East, and the Iberian Peninsula. Spanish priests, settlers and engineers brought it to Central and South America in the 19th century. These chain-wells (known in Spanish as *gálerias*) survive in Tehuacán, Parras, Puebla and Jalisco. Unfortunately, the idea never percolated north to the American West despite the region's abundant use of other Middle Eastern water technologies including *noria* (waterwheels) and the *shadoof* (in Mexico, the *bimbalete*), a fulcrum device for raising water that is common in the Southwest and is probably much older than chain-wells.

14. John Wesley Powell, *Report on the Land of the Arid Region of the United States.* (Washington, DC: US Government Printing Office, 1879): 6.

15. Wallace Stegner, *Beyond the 100th Meridian.* (Boston: Houghton Mifflin, 1954): 343, cited in David Sheridan, *Desertification of the United States.* (Washington, DC: Council on Environmental Quality, 1981): 6.

16. Donald E. Green, *Land of the Underground Rain: Irrigation on the Texas High Plains, 1910–1970.* (Austin: University of Texas Press, 1973): 15–29. This is a wonderfully readable work of painstaking scholarship that is unfortunately out of print. I can't recommend it enough, but it is a challenge to find a copy outside of the Southwest. See also, David Sheridan, *Desertification of the United States.* (Washington, DC: Council on Environmental Quality, 1981): 7.

17. Frank and Deborah Popper, "The Great Plains: From Dust to Dust," *Planning Magazine* Dec 1987. Retrieved online June 8, 2008 at planning.org.

18. Ibid.

19. Ibid.

20. Matthews, *Where the Buffalo Roam:* 18.

21. Ibid.

22. Midkiff, *Not a Drop:* 22.

23. Ibid., 30.

24. Associated Press, "Ghost Towning Steeped in Mysteries of Old West," retrieved on July 27, 2008 at cnn.com.

25. Edmund C. Jaeger, *The North American Deserts.* (Stanford: Stanford University Press, 1957): 3–12.

26. Matthews, *Where the Buffalo Roam:* 30.

27. Matt Moline, "Land Use Ideas Resurface," *Topeka Capital Journal* Feb 2, 2004. Retrieved online June 8, 2008 at cjonline.com.

28. Matthews, *Where the Buffalo Roam:* 29.

29. Michael Soulé, Reed Noss, "Rewilding and Biodiversity," *Wild Earth* 8 (3) 1998: 18–28. For more about the controversial history of "rewilding" see William Stoltenburg, *Where the Wild Things Were: Life, Death, and Ecological Wreckage in a Land of Vanishing Predators.* (New York: Bloomsbury USA, 2008): 174–83.

30. Cited in Lester R. Brown, *Plan B 2.0: Rescuing a Planet under Stress and a Civilization in Trouble.* (New York: W. W. Norton, 2006): 72.

31. William Cline, *Global Warming and Agriculture: Estimates by Country*. (Washington, DC: Center for Global Development, 2007): 47, 97. This is a nearly impenetrable book with useful but very hard-to-access information. That really is too bad because it assembles a great deal of really useful information about the dismal future of international agriculture under the coming climate changes.

32. IPCC WGII, Ch 17; 17.1; 17.2.2, cited in Gabrielle Walker and Sir David King, *The Hot Topic*. (London: Bloomsbury, 2008): 62.

33. Jennifer Stenhauer, "Water-Starved California Slows Development," June 7, 2008. Retrieved on June 7 from nytimes.com.

34. Broecker and Kunzig, *Fixing Climate*: 185.

35. A. L. Westerling et al., "Warming and Earlier Spring Increase Western U.S. Forest Wildfire Activity," *Science* 313, 940 (2006); published online July 6, 2006.

36. Marc Reisner, *Cadillac Desert*, revised edition. (New York: Penguin, 1993): 131.

37. Midkiff, *Not a Drop to Drink*: 5.

38. Ibid., Foreword by Robert Kennedy Jr., xiv.

39. Brian Johnson, ed., *Encyclopedia of Global Warming Science and Technology*. Vol 1 (Santa Barbara, CA: ABC-CLIO, 2009): 276, see entries for "Glacial Retreat" and especially "Glacier National Park"; see also David Trent et al., eds., *Geology and the Environment*. (Stamford, CT: Cengage Learning, 2010): 363.

40. Felicity Barringer, "In California, What Price Water?" *New York Times* Feb 28, 2013. Retrieved on Apr 19, 2013 from nytimes.com. See also Bettina Box, "Seawater Desalination Plant Might Be Just a Drop in the Bucket," *L.A. Times* Feb 17, 2013. Retrieved on Apr 15, 2013 from articles.latimes.com.

41. Ibid., 136–40; see also Terry Rodger, "San Onofre Desalination Plant Study Authorized," *San Diego Union Tribune* Oct 28, 2005.

42. Deborah Schoch, "L.A. Country Hopes To Fend Off Drought with Cloud Seeding Program," *Los Angeles Times*. June 16, 2008. Retrieved on June 16 at latimes.com. Anjali Athavaley, "Sewer to Spigot: Recycled Water," *Wall St. Journal* May 15, 2008. Retrieved June 6, 2008 from wsj.com.

43. Ibid., (*Sewer to Spigot*).

44. A. L. Westerling et al., "Warming and Earlier Spring Increase

Western U.S. Forest Wildfire Activity," *Science* 313, 940 (2006); published online July 6, 2006; Stephen W. Running, "Is Global Warming Causing More, Larger Wildfires?" *Science* 18, 313, 5789, Aug 18, 2006: 927–28.

45. Bettina Boxall, Julie Cart, "As Wildfires Get Wilder, the Costs of Fighting Them are Untamed," *Los Angeles Times* July 27, 2008. Retrieved on July 27 from latimes.com.

46. Vanessa Colón, "Farmworkers Leaving the Valley in Search of Jobs," *Fresno Bee* Sep 2, 2008, retrieved online on Sep 9, 2008.

47. Nelson (1918): 400; cited in Charles Taylor Vorhies, *Life of the Kangaroo Rat*. US Dept. of Agriculture Bulletin # 1091. One person who writes lovingly about kangaroo rats is Ann Zwinger, see *Wine in the Rock: The Canyonlands of Southeastern Utah*. (Phoenix: University of Arizona Press, 1986): 223–25.

Chapter 4

1. Gabrielle Walker, Sir David King, *The Hot Topic: How to Tackle Global Warming and Still Keep the Lights On*. (London: Bloomsbury, 2008): 213.

2. Ibid., 98.

3. IPCC, "Summary for Policymakers: Sec. 1 3. Projected Climate Change and Its Impacts," 2007: Table Spm. 1.

4. Julienne Stroeve et al., "Arctic Ice Sheet Decline: Faster than Forecast," *Geophysical Research Letters* Vol 34, 2007; Martin Redfern, "Antarctic Glaciers Surge to Ocean," BBC.com. Feb 25, 2008, available online at news.bbc.co.uk, retrieved Feb 25, 2008; Dr. Julian Scott in personal e-mail message Feb 2008.

5. Dr. Peter Noerdlinger, a professor at St. Mary's University in Nova Scotia, Canada. In a paper titled "The Melting of Floating Ice Will Raise the Ocean Level" submitted to *Geophysical Journal International*, Noerdlinger demonstrates that melt water from sea ice and floating ice shelves could add 2.6% more water to the ocean than the water displaced by the ice, or the equivalent of approximately 4 centimeters (1.57 inches) of sea level rise. The common misconception that floating ice won't increase sea level when it melts occurs because the difference in density between freshwater and saltwater is not taken into consideration. Archimedes's Principle states that an object immersed in a fluid is buoyed up

by a force equal to the weight of the fluid it displaces. However, Noerdlinger notes that because freshwater is not as dense as saltwater, freshwater actually has greater volume than an equivalent weight of saltwater. Thus, when freshwater ice melts in the ocean, it contributes a greater volume of melt water than it originally displaced.

6. Robin E. Bell, "The Unquiet Ice," *Scientific American* Feb 2008: 60–67.

7. California Climate Change Center, "The Impacts of Sea Level Rise on the California Coast," May 2009. Downloaded on Sep 27, 2012 from pacinst.org.

8. R. J. Rowley et al., "Risk of Rising Sea Level to Population and Land Area," *EOS: Transactions of the American Geophysical Union* Vol 88 (927) Feb 2007: 105–16, see table 1.

9. Asbury H. Sallenger Jr. et al., "Hotspot of Accelerated Sea-Level Rise on the Atlantic Coast of North America," *Nature Climate Change* June 24, 2012. Downloaded on Sep 27, 2012 from nature.com.

10. M. D. Powell, T. A. Reinhold, "Tropical Cyclone Destructive Potential by Integrated Kinetic Energy (IKE)," *BAMS* Apr 2007: 518.

11. Kerry Emanuel, "Increasing Destructiveness of Tropical Cyclones Over the Past 30 Years," *Nature* Vol 436, Aug 4, 2005.

12. Alexandra Bernardes Pezza, Ian Simmond, "The First South Atlantic Hurricane: Unprecedented Blocking, Low Sheer and Climate Change," *Geophysical Research Letters* Vol 32, Aug 15, 2005.

13. Mark Fischetti, "Drowning New Orleans," *Scientific American* Oct 1, 2001.

14. John McQuaid, Mark Schleifstein, *Path of Destruction: The Devastation of New Orleans and the Coming Age of Superstorms.* (New York: Little Brown, 2006): 125–26.

15. In John McQuaid, Mark Schleifstein, "The Big One," *Times-Picayune* June 24, 2002.

16. Ivor Van Heerden, *The Storm: What Went Wrong and Why During Hurricane Katrina: The Inside Story of One Louisiana Scientist.* (New York: Viking, 2006): 50; *Hurricanes* are a family of mixed drinks resembling daiquiris that combine lime juice, passion fruit syrup, and rum. They originated in WWII, at Pat O'Brien's Tavern in New Orleans and are properly served in a "Hurricane

Glass," so called because it is tall and has the same fluting as a hurricane lamp. Originally, they were made with surplus rum and given away free to sailors.

17. Ibid., 133.
18. In McQuaid and Schleifstein, "The Big One."
19. Van Heerden, *The Storm*: 50.
20. Timothy J. Hanley et al., "Families and Hurricane Response: Evacuation, Separation and the Emotional Toll of Hurricane Katrina," in David Brusma et al., eds., *The Sociology of Katrina: Perspectives on a Modern Catastrophe*. (New York: Rowman and Littlefield, 2007): 72.
21. Alan Berube, Steven Raphael, "Access to Cars in New Orleans," Brookings Institute, brookings.edu, Sep 15, 2005.
22. John Harlow, "Teenager Snatches Bus to Save Dozens," *London Times* Sep 4, 2005. Retrieved online at timeonline.co.uk, Feb 13, 2008.
23. Salathea Bryant, Cynthai Leonor Garza, "School Bus Commandeered by Renegade Refugees First to Arrive at Astrodome," *Houston Chronicle* Sep 1, 2005. Retrieved online at chro.com. Accessed on Feb 13, 2008.
24. Ibid.
25. Michael Lewis, "Wading towards Home," *New York Times* nytimes.com, Oct 9, 2005. Accessed Jan 11, 2008.
26. Bryant and Garza, "School Bus Commandeered." Accessed online at chro.com on Oct 2, 2012.
27. Douglas Brinkley, *The Great Deluge: Hurricane Katrina, New Orleans and the Mississippi Gulf Coast*. (New York: William Morrow, 2006): 515–16.
28. Chris Wood, *Dry Spring: The Coming Water Crisis of North America*. (Vancouver: Greystone, 2008): 79.
29. Lewis, "Wading towards Home."
30. Ibid.
31. This is an incomplete record of my hastily scratched notes following the telephone conversation I had with Jabar Gibson on October 10, 2012. We spoke twice over the phone. But he called that first time completely unexpectedly after receiving a letter that provided my contact information and described this book. I mailed the manuscript pages describing his Katrina experience to him the following morning, but I never got a response.

32. John M. Barry, *Rising Tide: The Great Mississippi Flood of 1927 and How It Changed America*. (New York: Simon and Schuster, 1997). An incorrect version of the incident with the calliope is described in Evan Thomas, "The Lost City," *Newsweek* Sep 12, 2005: 56. Barry's original source is listed as the "oral history of Ernest Waldauer" archived in the Mississippi Department of Archives and History at Jackson. Thomas seems to have made a composite of several incidents described individually in *Rising Tide*.

33. "Calling Survivors Refugees Stirs Debate," MSNBC.com, Sep 7, 2005. Accessed Jan 27, 2008 at msnbc.com.

34. William Safire, "Katrina Words," On Language column, *New York Times Magazine* Sep 18, 2005. Accessed Feb 16, 2008 at nytimes .com.

35. "Calling Survivors Refugees Stirs Debate."

36. Van Heerden, *The Storm*: 66.

37. German Advisory Council on Climate Change, *Climate Change as a Security Risk*. (London: Earthscan, 2008): 120.

38. Jed Horne, *Breach of Faith: Hurricane Katrina and the Near Death of a Great American City*. (New York: Random House, 2006): 17.

39. Cited in Lester R. Brown's *Full Planet, Empty Plates: The Geopolitics of Food Scarcity*. (New York: W. W. Norton, 2012): 89. No source given for: "if the Greenland ice sheet were to melt entirely, it would raise sea level 23 feet. The latest projection show sea level rising by 6 feet this century."

40. Ibid., 88.

41. Susmita Dasgupta, "The Impact of Sea Level Rise on Developing Countries: A Comparative Analysis," *Policy Research Working Paper no. WPS 4136*: 44, econ.worldbank.org. Retrieved online on Feb 8, 2008.

42. Julie Angus, *Rowboat in a Hurricane: My Amazing Journey across a Changing Atlantic Ocean*. (Vancouver: Greystone, 2008): 101–2.

43. R. A. Scotti, *Sudden Sea: The Great Hurricane of 1938*. (New York: Little Brown, 2003): 209, 226–28.

44. Vivien Gornitz et al., "Impacts of Sea Level Rise in the New York City Metropolitan Area," *Global and Planetary Change* 32, 2002: 61–88; but for a much clearer and more readable account see Mark Lynas, *Six Degrees: Our Future on a Hotter Planet*. (London: Fourth Estate, 2007): 157–59.

45. Janine Bloomfield, *Hot Nights in the City: Global Warming, Sea*

Level Rise and the New York Metropolitan Region. (Environmental Defense Fund, 1999): 44.

46. Bloomfield's remarks are published on the Internet and can be retrieved at climatehotmap.org.

47. Mark Lynas, *Six Degrees: Our Future on a Hotter Planet.* (London: Fourth Estate, 2007): 159.

48. Rose George, *The Big Necessity: The Unmentionable World of Human Waste and Why it Matters.* (New York: Metropolitan Books, 2008): 17.

49. Asbury H. Sallenger Jr. et al., "Hotspot of Accelerated Sea-Level Rise on the Atlantic Coast of North America," *Nature Climate Change* June 24, 2012. Retrieved Oct 2012 from nature.com.

50. Ibid.

Chapter 5

1. Anna Smolchenko, "Russia Faces 15 Billion Dollars in Heatwave Losses," Aug 10, 2010, retrieved online on Aug 10 from google .com/hosted news/AFP.

2. Edward Kohn, *Hot Time in the Old Town: The Great Heat Wave of 1896.* (New York: Basic Books, 2010): 74.

3. Bryan Walsh, "Parched Earth," *Time Magazine* Aug 22, 2011. Downloaded on Oct 27, 2012 from time.com.

4. Paul Stenquist, "A Record Number of Children Are Dying in Hot Cars," *New York Times* Aug 9, 2010. Retrieved online on Aug 10, 2010 at wheels.blogs.nytimes.com.

5. Kohn, *Hot Time in the Old Town*: 191–92.

6. Maria Blunt, "The Fate of the Georgiana," *Scribner's Magazine* Vol 4 (2) Aug 1888: 222.

7. James Charles Van Dyke, *The New New York: A Commentary on the Place and the People.* (New York: Macmillan, 1909): 56.

8. Staff Correspondence: Huntington, *The Congregationalist.* Aug 20, 1896: 81, 34. My deepest thanks, incidentally, to Emily Howie at the Library of Congress for suggesting this source, and then for patiently printing and mailing it to me in distant Vancouver, Canada, where there is no research library with sufficient resources to access ProQuest's APS online service where the document is archived.

9. Steven Smith, "Vegetation a Remedy for the Summer Heat of Cities," *Appleton's Popular Science Magazine* 31, Feb 1899: 440.

Daily horse excrement figures are derived from the annual figures cited in David Suzuki and David Boyd's *David Suzuki's Green Guide*. (Vancouver: Douglas and McIntyre, 2008): 71. If David puts his name to it, it's good enough for me.

10. Gail Cooper, *Air Conditioning America: Engineers and the Controlled Environment*. (Baltimore: Johns Hopkins, 1998): 8–17.

11. U.S. Bureau of Animal Husbandry, *Special Report on Diseases of the Horse*. (Washington, DC: Government Printing Office, 1903): 199.

12. Kohn, *Hot Time in the Old Town*: 87–88.

13. William Dean Howells, "The Silver Wedding Journey," (Part 1) *Harpers* Vol 98 (584) Jan 1899: 200.

14. Kohn, *Hot Time in the Old Town*: 91 (no source given).

15. Kohn, *Hot Time in the Old Town*: ix, 184. But see also: Staff Correspondence: Huntington, *The Congregationalist* Aug 20, 1896: 81, 34.

16. E. Wong, "Green Roofs and the U. S. Environmental Protection Agencies Heat Island Reduction Initiative," in *Proceedings of the Third Annual International Green Roofs Conference: Greening Rooftops for Sustainable Communities*. (Washington, DC) May 2003; cited in Nigel Dunnet, Noel Kingsbury, *Planting Green Roofs and Living Walls*. (Portland: Timber Press, 2008): 126.

17. Lisa Gartland, *Heat Islands: Understanding and Mitigating Heat in Urban Areas*. (London: Earthscan, 2008): 6. This is the only book in existence about the phenomenon. Every library should have a copy.

18. Gartland, *Heat Islands*: 8.

19. "The Recent Heatwave," *Scientific American* Aug 22, 1896; Vol LXXV, no. 8, via Proquest's APS Online, 166. Once again, I am deeply indebted to the Library of Congress and especially to librarian Emily Howie for drawing my attention to this article.

20. Ibid., 166.

21. This figure is from a *New York Times* article of the period, and will probably show up if you search keywords for "Barge Captain," "1258 dead horses," or "Barren Island," in ProQuest Historical Newspapers. I've lost the bloody reference, and am confessing my carelessness to the world instead of spending the day looking for the exact piece. Perhaps you could just trust me on this one. If you can, then I thank you.

22. Kohn, *Hot Time in the Old Town*, Appendix A: Death Certificates Filed: 261.

23. F. P. Ellis, "Mortality from Heat Illness and Heat-Aggravated Illness in the United States," *Environmental Research* 5, 1972: 1–58.

24. William B. Meyer, *Americans and Their Weather*. (Oxford: Oxford University Press, 2000): 81. This is a really enjoyable and informative book.

25. Ibid., 82.

26. Ibid., 82, note 165, but see also Henry Collins Brown, *In the Golden Nineties*. (Hastings on Hudson: Valentine's Manual, 1928): 353.

27. Kohn, *Hot Time in the Old Town*: 72.

28. Lawrence Jeffrey Epstein, *At the Edge of a Dream: The Story of Jewish Immigrants on New York's Lower East Side, 1880–1920*. (New York: Wiley, 2007): 41.

29. Kohn, *Hot Time in the Old Town*: 7. (No source for Roosevelt's quote is cited.)

30. Ibid., 57.

31. Jacob A. Riis, *How the Other Half Lives: Studies among the Tenements of New York City*. (New York: Dover Publications, 1971): 126; cited in Meyer, 124.

32. Epstein, *At the Edge of a Dream*: 57.

33. Kohn, *Hot Time in the Old Town*: 11–12.

34. Editor's Easy Chair, *Harper's* Vol 57, issue 341, Oct 1878, 784.

35. "The Recent Heat Wave," *Scientific American* Aug 22, 1896, Vol LXXV, no. 9: 166.

36. Wallace Broecker, Robert Kunzig, *Fixing Climate: What Past Climate Changes Reveal about the Current Threat—And How to Counter It*. (New York: Hill and Wang, 2008); 176–79.

37. Kate Linthicum, Andrew Blankstein, "Mummified Remains of Two Babies Wrapped in 1930s Newspapers Found," *Los Angeles Times* Aug 19, 2010. Accessed Aug 19, 2010 at latimes.com.

38. Clarence A. Mills, *Climate Makes the Man*. (New York: Harper and Brothers, 1942): 67.

39. Kohn, *Hot Time in the Old Town*: 57.

40. Jane E. Dematte et al., "Near-Fatal Heat Stroke during the 1995 Heat Wave in Chicago," *Annals of Internal Medicine* Vol 129 no. 3, Aug 1, 1998.

41. George Shattuck, Margaret Hilferty, "Causes of Deaths From

Heat in Massachusetts," *New England Journal of Medicine* Vol 209 no. 7, Aug 17, 1933, 319–29.

42. Eric Klinenberg, *Heat Wave: A Social History of Disaster in Chicago.* (Chicago: University of Chicago Press, 2002): 27.

43. Andrew Serl et al., "When Can We Expect Extremely High Surface Temperatures," *Geophysical Research Letters* Vol 35, 2008. Unpaginated courtesy copy from Professor Serl, who generously responded extensively to my questions about extreme temperatures in the continental United States of the future.

44. Klinenberg, *Heat Wave*: 26.

45. Ibid.

46. Fran Spielman and Mary Mitchell, "City Ignored Emergency Plan," *Chicago Sun-Times* July 18, 1995: 1; cited in Klinenberg, *Heat Wave*: 169.

47. Klinenberg, *Heat Wave*: 169.

48. *New York Times* July 11, 1936: 2.

49. Klinenberg, *Heat Wave*: 200.

50. Kohn, *Hot Time in the Old Town*: x.

51. Unattributed quote cited in Klinenberg, *Heat Wave*: 173.

52. Klinenberg, *Heat Wave*: 169.

53. Gilbert Jimenez, Alex Rodríguez, "Death Toll Climbs to 179," *Chicago Sun-Times* July 18, 1995: 6; John Kass, "Daley Aides Try to Deflect Heat Criticism," *Chicago Tribune* July 18, 1995: sec 2 p 1; Fran Spielman, Mary Mitchell, "The Shocking Toll: 376," *Chicago Sun-Times* July 19, 1995: 1, 9: cited in Klinenberg, *Heat Wave*: 169.

54. John Kass, "Daley Aides Try to Deflect Heat Criticism," *Chicago Tribune* July 18, 1995: sec 2 p 1: cited in Klinenberg, *Heat Wave*: 172.

55. Fran Spielman, Mary Mitchell, "City Ignored Emergency Plan," *Chicago Sun-Times* July 18, 1995. Cited in *Heat Wave*: 172.

56. Sharon Cotliar, "Count Isn't Overblown Medical Examiner Insists," *Chicago Sun-Times* July 19, 1995: 8. Cited in *Heat Wave*: 172.

57. Ibid.; Joel Kapland, Sharman Stein, "City Deaths in Heat Wave Triple Normal," *Chicago Tribune* July 20, 1995, sec 1 p 1, 10. Cited in *Heat Wave*: 172.

58. In Klinenberg, *Heat Wave*: 173.

59. Ibid., 31.

60. Ibid., 29.

61. Ibid., 174.

62. Mark Lynas, *Six Degrees: Our Future on a Hotter Planet.* (London: Fourth Estate, 2007): 200.
63. Nigel Dunnet, Noel Kingsbury, *Planting Green Roofs and Living Walls.* (Portland: Timber Press, 2008): 78.
64 Andrew Serl et al., "When Can We Expect Extremely High Surface Temperatures," *Geophysical Research Letters* Vol 35, 2008.
65. Personal email from Professor Serl; but see also L. Ruby Leung, "Mid-Century Ensemble Regional Climate Change Scenarios for the Western United States," *Climatic Change* 62, 2004: 75–113.
66. Oliver Deschenes, Michael Greenstone, "Climate Change, Mortality and Adaptation: Evidence from Annual Fluctuations in Weather in the U.S.," *Massachusetts Institute of Technology, Department of Economics Working Paper* 07–19, July 21, 2007.
67. Mark Lynas, *Six Degrees:* 70.
68. C. Schar et al., "The Role of Increasing Temperature Variability in Summer Heatwaves," *Nature* 427: 332–36. See also Lynas, *Six Degrees:* 66.

Chapter 6

1. Scandinavians still commonly use such "plant houses" to grow vegetables and herbs during their northern winters. The Swedish word for plant (*väx*) shares its root with the Swedish words for *grow, wake* and *beautiful*, which perhaps indicates the impact of harsh, long, northern winters on reluctantly frigid Swedes. Until the turn of the century, these structures were uncommon in most of the English-speaking world although they existed in England, where the name *växthaus* was translated as "greenhouse." They only became known in North America after the Swedes took advantage of the declining cost of Atlantic crossings and the cheap land made available in the north-most border states of the Midwest, from Illinois to Montana.

   Before World War I, the potential for a more prosperous life in America attracted many Scandinavians who were disillusioned by the roller coaster of their domestic economy. By the 1920s, 20% of the population of Sweden had come to America, making it their new home. (Hans Norman, "The Causes of Emigration," in Runblom and Norman, *From Sweden to America* (1976): 149–64.) As they built houses and gardens in the United States, the utility of their greenhouses began to attract interest among their

American neighbors. In 1907, Paul N. Hasluck published *Greenhouse and Conservatory Construction and Heating*, which popularized the Swedes' method of winter gardening throughout the northern United States.

2. Associated Press, "UN Says Greenhouse Gases at Record High in 2011," Nov 20, 2012, weather.com. Accessed May 6, 2013.

3. Bill McKibben, *eaarth: Making a Life on a Tough New Planet.* (New York: Times Books, 2010): 20–22.

4. Michael Weber, "Will Drought Cause the Next Blackout?" *New York Times* Op-ed, July 23, 2012.

5. Doyle Rice, "2012 Off to Furious Start in Tornadoes," *USA Today* Jan 31, 2012. Downloaded on Oct 11, 2012 from usatoday.com

6. "Storm, Tornadoes, Damage Missouri, Illinois, Kansas; Kill 9," *Christian Science Monitor* Feb 29, 2012. Downloaded on Oct 11, 2012 from csmonitor.com.

7. Brian McNoldy, "Superstorm Sandy Packed More Total Energy than Hurricane Katrina at Landfall," *Washington Post* blogs. Nov 2, 2012.

8. Mikyong Shin et al., "Prevalence of Down Syndrome among Children and Adolescents in 10 Regions of the United States," *Pediatrics* Vol 124 no. 6, Dec 1, 2009: 1565–71.

9. Alex Prud'homme, *The Ripple Effect: The Fate of Fresh Water in the 21st Century.* (New York: Simon and Schuster, 2011): 261.

10. James Geisen, *Boll Weevil Blues: Myth and Power in the American South.* (Chicago: University of Chicago Press, 2011): 98.

11. Joe Trotter, *The Great Migration in Historical Perspective.* (Bloomington: Indiana University Press, 1991): 5.

12. Geisen, *Boll Weevil Blues:* 98.

13. CNN Money, "Older Americans Are 47 Times Richer than Young," Nov 28, 2011.

14. AP story, "US Jobs Gap between Young and Old is Widest Ever," finance.yahoo.com, Feb 9, 2012.

15. Laurence Kotlikoff, Scott Burns, *The Clash of Generations: Saving Ourselves, Our Kids and Our Economy.* (Cambridge, MA: MIT Press, 2012): 229.

16. Thomas C. Peterson et al., eds., "Explaining Extreme Events of 2100 from a Climate Perspective," *Bulletin of the American Meteorological Society* (BAMS) July 2012: 1052.

17 *Weekly Weather and Crop Bulletin* US Department of Commerce; US Dept. of Agriculture. Vol 99 no. 32, Aug 7, 2012: 2.

18. L. A. Urrea, "Ghosts of America: Memories of the Present and Other Signs of the Times," *Orion*, Jan 2013. Accessed Jan 17, 2013 at orionmagazine.org.

19. Bryan Walsh, "Parched Earth," *Time Magazine* Aug 22, 2011.

20. "Global Weather Extremes Summary," blog posts Weather Underground, wunderground.com.

21. Brendan Choat et al., "Global Convergence in the Vulnerability of Forests to Drought," *Nature* Nov 29, 2012: 752–55.

22. Christopher Joyce, "Is It Too Late to Defuse the Danger of Mega-Fires?" Aug 24, 2012, npr.org.

23. Jim Forsyth, "Texas Drought Kills As Many As Half a Billion Trees," reuters.com, Dec 20, 2011.

24. Mike Agresta, "The 'D' Word: Drought," *The Alcalde* texasexes.org, Apr 30, 2012. Accessed Oct 14, 2012.

25. Walter Van Tilburg Clark, *The City of the Trembling Leaves*. (New York: Random House, 1945): 546.

26. "Historic Dust Storm Sweeps Arizona, Turns Day to Night," Reuters, July 6, 2011.

27. Sharon Rowen, Beverly Bryant, "Dust Storm 2012 Close I-35 Thursday," *Ponca City News* Oct 19, 2012: 1. Downloaded on Apr 19, 2013 from assetsmediaspanonline.com.

28. Beverly Bryant of the *Ponca City News*, in conversation with author.

29. Bryan Walsh, "Parched Earth," *Time Magazine* Aug 22, 2011.

30. Michael Weber, "Will Drought Cause the Next Blackout?" *New York Times* Op-ed, July 23, 2012.

31. Ibid.

32. "UN Calls on Nations to Adopt Urgent Drought Policies," *The Guardian* Aug 21, 2012.

33. Walsh, "Parched Earth."

34. Larry Elliott, "US Drought Will Lead to Inflation and Higher Food Prices," *The Guardian* Aug 20, 2012.

35. Martin Hoerling, the scientist who first connected the Texas 2011 drought to climate change, now believes the more severe 2012 drought is not similarly connected. In M. Hoerling et al., "An Interpretation of the Origins of the 2012 Central Great Plains

Drought," *NIDIS* Apr 11, 2013. Downloaded on Apr 12 from cpo.noaa.gov. Noted meteorologist Kevin Trenbreth, however, disagrees, noting that Hoerling's new models fail to take into account the decline in local American snow and ice cover which contributes to reduced river flows, drier soil and the ever-lowering levels of aquifers, a factor which continues to influence the ongoing drought in the Southwest in 2013; Suzanne Goldberg, "Climate Change Did Not Cause 2012 US Drought, Says Government Report," *The Guardian*. Downloaded on Apr 12 from guardian.co.uk.

36. A diesel engine works at about 30–35% greater efficiency than a normal gas-powered vehicle, but even so, the average big-rig gets between 4.5 and 7 mpg on a flat road, and fuel is only one of many costs for freight shipped by truck. Consequently, trucking freight is a much more expensive option than shipping by barge.

37. John Yang, "Drought Sends Mississippi into 'Uncharted Territory,'" nbcnews.com, Nov 30, 2012.

38. Alan Bjerga, "Drought-Parched Mississippi River is Halting Barges," bloomberg.com, Nov 27, 2012.

39. Ibid.

40. Linda McMaken, "How the Severe Drought Will Affect Americans," Aug 22, 2012. Downloaded on Apr 19, 2013 from investopedia.com.

41. Yang, "Drought."

42. Bjerga, "Drought-Parched Mississippi."

43. Hilary Hylton, "'Forget Irene: The Drought in Texas Is the Catastrophe That Could Really Hurt,'" time.com, Aug 31, 2011.

44. Brad Plumer, "What We Know about Climate Change and Drought," *Washington Post* July 24, 2012.

45. Ibid.

46. Thomas C. Peterson et al., eds., "Explaining Extreme Events of 2100 from a Climate Perspective," *Bulletin of the American Meteorological Society* (BAMS) July 2012: 1053.

47. Their names should be mentioned here: Frenchman Jacques Anquetil; the great Belgian cyclist Eddy Merckx, who zipped down a road past me in the Pyrenees when I was 16; and the other great Frenchman, Bernard Hinault. Despite a superhuman effort, Hinault lost the 1986 race to its first non-European winner, the American anti-doping activist, Greg LeMond, whose reputation

was destroyed when doping allegations were made against him by Lance Armstrong.

48. Brad Knickerbocker, "Hurricane Sandy Blows Climate Change Back into the Presidential Race," *Christian Science Monitor* Nov 3, 2012.

49. Both quotes are from: C. B. Field et al., eds., "2012: Summary for Policymakers," in *Managing the Risks of Extreme Events and Disasters to Advance Climate Change Adaptation. A Special Report of Working Groups I and II of the Intergovernmental Panel on Climate Change*. (New York: Cambridge University Press, 2012): 5 and 9.

50. Peterson et al., "Explaining Extreme Events of 2011": 1052–53.

51. Nikolas Christidis et al., "The Role of Human Activity in the Recent Warming of Extremely Warm Daytime Temperatures," *BAMS* Vol 24, 2011: 1922.

52. Janet Larson, "Plan B Updates: Setting the Record Straight: More than 52,000 Europeans Died from Heat in Summer 2003," July 28, 2006. Downloaded on Oct 28, 2012 from earth-policy.org; World Health Organization, downloaded on Jan 15, 2012 from euro.who.int.

53. Earth Policy Institute quoted in Shaoni Bhattacharya, "European Heatwave Caused 35,000 Deaths," *New Scientist* Oct 10, 2003. Downloaded on Oct 28, 2012 from newscientist.com.

54. C. B. Field et al., eds., "2012: Summary for Policymakers": 5.

55. Michael Wehner et al., "Projections of Future Drought in the Continental United States and Mexico," *BAMS* Apr 2011: 1359–77.

56. David Easterling, Michael Wehner, "Is the Climate Warming or Cooling?" *Geophysical Research Letters.* Article first published online Apr 25, 2009.

57. R. Seager, "Making a Bad Situation Worse: Climate Change and Human-Induced Aridification of Southwestern North America," *California Department of Water Resources California Drought Update* June 2008: 12.

58. C. B. Field et al., eds., "2012: Summary for Policymakers": 12.

Chapter 7

1. Suzanne Goldenberg, "Sandy Forces New York to Consider All Its Options in Effort to Make City Safe," *The Guardian* Nov 2, 2012. Accessed Jan 29, 2013 at guardian.co.uk.

2. Jim Van Anglen, "Mobile Damaged by Tornado: 23,500 without Power," *Gadsden Times* Dec 25, 2012. Downloaded from gadsden times.com; see also Janet McConnaughey, Jim Van Anglen, "Christmas Day Storms Blamed for Three Deaths," *Washington Times* Dec 25, 2012. Downloaded from washingtontimes.com.

3. Bonnie Schneider, *Extreme Weather: A Guide to Surviving…Natural Disasters.* (New York: MacMillan, 2012): 24–25.

4. A.A. Justice, "Seeing the Inside of a Tornado," *Monthly Weather Review* May 1930: 205.

5. Lieut. John Park Finley, *Tornadoes: What They Are and How to Observe Them.* (New York: Insurance Monitor Press, 1887).

6. Public Religion Research Institute, "Americans More Likely to Attribute Increasing Severe Weather to Climate Change, Not End Times." Accessed Dec 12, 2012 from publicreligion.org.

7. John Dickerson, "The Decline of Evangelical America," *New York Times* Opinion column, Dec 16, 2012: 5.

8. Munich Re, "Severe Weather in North America," cited in Elizabeth Kolbert, "Watching Sandy, Ignoring Climate Change," *The New Yorker* Oct 29, 2012.

9. A.M. Best Company, Inc., *Tornado Losses Approach Those of Hurricanes.* (Oldwick, NJ: A. M. Best Co., 2008): 1.

10. Ibid., 7.

11. A wonderful American word describing the phenomenon of deliberately created confusion.

12. Tim Samaras, Stefan Bechtel, *Tornado Hunter.* (Washington, DC: National Geographic Books, 2009): unpaginated Kindle edition, Ch 13: Global Warming and Tornadoes.

13. Liz Robbins, "Nine Killed as Tornado Rakes Oklahoma," *New York Times* Feb 12, 2009. A 22. Available at nytimes.com.

14. Ibid.

15. Ibid.

16. Chris Mooney, "Is Climate Change Causing an Upsurge in US Tornados?" *New Scientist* Aug 1, 2008.

17. Christine Archer, Ken Caldiera, "Historical Trends in the Jets Streams," *Geophysical Research Letters* Vol 35 (8) 2008.

18. Q. Fu et al., "Enhanced Mid-Latitude Tropospheric Warming in Satellite Measurements," *Science* 312 (2006): 1179; Y. Hu, Q. Fu, "Observed Poleward Expansion of the Hadley Circulation since

1979," *Atmospheric Chemistry Physicals Discussions* 7 (2007): 9364, 9367.

19. Archer and Caldiera, "Historical Trends in the Jets Streams."

20. Samaras and Bechtel, *Tornado Hunter.*

21. Jon Pareles, "Basking in the Sun, Though Wary of a Storm," *New York Times* Feb 5, 2008: E1.

22. Howard Bluestein, *Tornado Alley: Monster Storms of the Great Plains.* (New York: Oxford University Press, 1999): unpaginated Kindle edition.

23. National Oceanic and Atmospheric Administration, "Tornadoes: Nature's Most Violent Storms," nssl.noaa.gov, accessed Jan 12, 2012.

24. Bluestein, *Tornado Alley:* unpaginated Kindle edition.

25. Lieut. John Park Finley, *Tornadoes: What They Are and How to Observe Them.* Published in 1887 by the Insurance Monitor Press, NYC; first edition.

26. USGCR, *Draft Climate Assessment Report.* Revised Jan 11, 2013:39, ncadac.globalchange.gov, accessed Feb 10, 2013.

27. Thomas C. Peterson et al., eds., "Explaining Extreme Events of 2100 from a Climate Perspective," *BAMS* July 2012: 1041–67.

28. Sometimes Ibom was called Iya (Evil One). He is identified with eye symbolism and with incarnate evil. He hates human beings, who are the legitimate (grand)children of his father, Inyan. Tornado's younger brother, Gnas, is the Devil; he is similarly evil and appears as a compellingly beautiful young man who is mankind's great deceiver and destroyer. (In: Phil Hart, *The Book of Imaginary Indians: Ancient Traditions and Modern Caricatures in the White Man's Quest for Meaning.* [Lincoln, NE: iUniverse, 2008]: 87; but see also, James R. Walker, *Lakota Myth.* [Lincoln: University of Nebraska Press, 2011]: 315.) Tornadoes were especially threatening events to Plains Indians like the Arapaho, Cheyenne, Kiowa and Sioux. Tribal belief systems ascribe different narratives to a uniform pantheon of divinities, but in all variations Ibom is consistently humanity's perverse and evil enemy. His hatred of human beings derives from his envy of mankind's legitimacy. When Inyan was still the most powerful god (long before he was seduced by Passion, Ibom's mother), he created the earth goddess Makha-akan (Makȟá-akáŋ) out of his own blood. And although this action made Inyan (Íŋyaŋ) very weak, he then

created miniature versions of Makha and her lover Skan (father sky), who would become the most powerful of the four gods. The miniatures Inyan created were human beings, his legitimate and best-loved children (or perhaps grandchildren). Ibom arrived by an unhappy accident well after Inyan created humans. Ibom's shape and color are a perverse parody of the human beings' white-colored tents named by the Lakota word *t'ipi* (meaning "they dwell," or dwelling). The Plains Indians often paint celestial or vision quest designs on the outer walls of their t'ipis. Many of these incorporated the blue and/or red blood colors that represent Inyan's blood from which they were created: but such colors also universally serve as a fundamental protection against the Evil Eye. Some designs, however, are simple geometries of black and white representing the cosmic opposition of good and evil, or of mankind and Unk's children, Ibom and the Devil. (Alan Dundes, "Wet and Dry and the Evil Eye: An Essay in Indo-European and Semitic Worldview," in *The Evil Eye: A Casebook.* [Madison: University of Wisconsin Press, 1981]; Paul Goble, *Tipi: Home of the Nomadic Buffalo Hunters.* [Bloomington: World Wisdom Books, 2007]). Moreover the t'ipi's portability represents the transhumance of their buffalo-hunting nomadic inhabitants. Quickly disassembled, a t'ipi is soon packed into a travois to be dragged behind the Lakota horses. These people did not see themselves as transient wanderers. They were always at home between their Earth Mother and Sky Father. Their houses came with them. Wherever they were on the Great Plains, they were as comfortable and self-sufficient as Ta-te, the wind-god (Paul Goble, *Tipi*).

29. "Extreme Weather Summary," wunderground.com, blog posts, 2011. But see also Jason Samenow, "Spring Extreme Weather Events in 2011 in U.S.: Historic and Record Setting," *Washington Post Weather Blog* June 16, 2011, washingtonpost.com.

30. Aldo Leopold, *A Sand County Almanac.* (New York: Oxford University Press, 1949): 153–54.

31. Michael Pearson, Phil Gast, Vivian Kuo, "High Winds, Tornado Trap Georgia Residents, Turn over Cars," cnn.com, accessed Jan 31, 2013; John Newland, Andrew Mach, "Tornado Rips through Georgia City as Storms Wreak Havoc in the South," usnews.com, accessed Jan 31, 2013.

32. Charlie LeDuff, *Detroit: An American Autopsy.* (New York: Penguin, 2013).

33. Dmitry Orlov, *The Five Stages of Collapse: A Survivor's Toolkit.* Feb 2013 (digital proof of ms).

34. Suzanne Goldenberg, "Sandy Forces New York to Consider All Its Options…," *The Guardian* Nov 2, 2012.

35. Climate Central, *Global Weirdness: Severe Storms, Deadly Heat Waves, Relentless Drought, Rising Seas, and the Weather of the Future.* (New York: Pantheon, 2012): 141–42.

36. The 1.6 million African American who participated in the Great Migration had an enormous impact on America, but this occurred when the United States had less than 100 million people.

37. John Heilprin, "UN Says Greenhouse Gases at Record High in 2011," *The Guardian* Nov 21, 2012; accessed Nov 21 from guardian.co.uk.

38. Kevin Trenberth, "What Role Did Climate Change Play in This Week's Massive Hurricane?" Oct 31, 2012. *The Scientist.* Downloaded from the-scientist.com.

39. Seth Borenstein, "Scientists Look at Climate Change: The Superstorm," news.yahoo.com, Oct 30, 2012. Accessed on Oct 30, 2012.

40. Ibid.

41. Peter Gleick, "The Very Real Threat of Sea Level Rise to the United States," HuffingtonPost.com. Downloaded on Sep 18, 2012 from huffingtonpost.com.

42. Neela Bannerjee, "Ice Sheet Melting Accounts for 20% of Sea Level Rise since 1992," *L.A. Times* Nov 29, 2012.

43. M.D. Powell, T.A. Reinhold, "Tropical Cyclone Destructive Potential by Integrated Kinetic Energy (IKE)," *BAMS* Apr 2007: 513–26.

44. Brian McNoldy, "Superstorm Sandy Packed More Total Energy than Hurricane Katrina at Landfall " *Washington Post* blogs. Nov 2, 2012.

45. Ibid.

46. William Westhoven, *Superstorm Sandy: A Diary in The Dark.* (CreateSpace Independent Publishing Platform, December 2, 2012): 5. Mr. Westhoven promised to donate all proceeds from this book to the nonprofit Hurricane Sandy New Jersey Relief Fund.

47. Damian Carrington, "Greenland and Antarctica Have Lost Four Trillion Tones of Ice in 20 Years," *The Guardian* Nov 29, 2012.

48. Bannerjee, "Ice Sheet Melting."

49. Seth Borenstein, "Scientists Look at Climate Change."

50. Charles Q. Choi, "Sea Level Rising Fast on U.S. East Coast," *National Geographic News* June 25, 2012. Downloaded from news.nationalgeographic.com.

51. Bannerjee, "Ice Sheet Melting."

52. Charles Q. Choi, "Sea Level Rise."

53. Powell and Reinhold, "Tropical Cyclone Destructive Potential."

54. Climate Central, *Global Weirdness: Severe Storms, Deadly Heat Waves, Relentless Drought, Rising Seas, and the Weather of the Future*. (New York: Pantheon, 2012): 141–42.

55. Goldenberg, "Sandy Forces New York to Consider All Its Options."

56. Kevin Trenberth, "What Role Did Climate Change Play?"

Chapter 8

1. William Carlos Williams, *In the American Grain*. (New York: New Directions Publications, 1990): 136.

2. Trapper, "Bert Blake" in Elliott Merrick's remarkable memoir *True North*. (North Ferrisburg, Vermont: Heron Dance Press, 2005 [original edition 1933]): 9.

3. IPCC (1996), in J.T. Houghton et al., "Climate Change 1995: The Science of Climate Change, Contribution of Working Group I to the Second Assessment Report of the Intergovernmental Panel on Climate Change," Cambridge University Press, 1995.

4. Alistair Doyle, Sven Ergiter, "U.N.'s Ban Seeks Strong Climate Pact, Fears Sea Level Rise," Reuters, Sep 3, 2009. Downloaded on May 4, 2013 from reuters.com.

5. S.L. Mitchell (1807) quoted in Hendrik Hartog, *Public Property and Private Power: The Corporation of the City of New York in American Law: 1730–1870*. (Ithaca, NY: Cornell University Press, 1983): 94.

6. Concerning the Loyalists, see W. Stewart Wallace, "The United Empire Loyalists: A Chronicle of the Great Migration," *Chronicles of Canada* Vol 13, 1914, Toronto (see especially Chapter 12, "The American Migration").

7. Concerning the war resisters, please note that the young men avoiding military service in Vietnam who resettled in Canada numbered only between 20,000 and 30,000, while the actual number of Americans—including wives, children and parents—who migrated at that time is much larger. See Douglas A. Ross, *In the Interests of Peace: Canada and Vietnam, 1954–73*. (University of Toronto Press, 1984), but there are many other fine books (and quite a bit of crap, too) written about this recent migration.

8. Nicholas Stern, *The Global Deal: Climate Change and The Creation of a New Era of Progress and Prosperity*. (New York: Public Affairs, 2009): 3.

9. Robert McLemen, "What If Climate Change Drives Americans Across Our Border?" *Globe and Mail* Feb 15, 2007. In .pdf format, generously provided by the author.

10. David Jones, *Empire of Dust: Settling and Abandoning the Prairie Dry Belt*. (Calgary: University of Calgary Press, 1987): 44.

11. Gerry Warner, "Disappearing Glacier at the Edge of Ralph's Kingdom," *Daily Bulletin* Kimberley, B.C., Nov 8, 2002: 4.

12. David Bly, "Farm Family Moves to Greener Pastures," *Calgary Herald* Oct 25, 2002: B5.

13. Gina Teel, "Alberta Soil Too Dry for Crops to Germinate," *Calgary Herald* Apr 9, 2002: E3.

14. Cited in Bernard Augustine de Voto, *The Course of Empire*. (Boston: Houghton Mifflin, 1980 [original edition 1952]): 211.

15. Stuart Logie, "Predicting Forest Flareups," *Montreal Gazette* Jan 25, 2003: B9; Margret Kopala, *Ottawa Citizen* Aug 30, 2006: B3.

16. Max Finkelstein, James Stone, *Paddling The Boreal Forest: Rediscovering A. P. Low*. (Toronto: Natural Heritage Books, 2004): 172.

17. Ibid., 93–94.

18. Especially in the western Yukon, the northeastern portion of the continental Northwest Territories, Nunavut, and all of the Arctic Archipelago.

19. Climate Change Impacts and Adaptation Program, *Climate Change Impacts and Adaptation: A Canadian Perspective* (National Resources Canada, 2004): 80.

20. M.D. Flannigan et al., "Future Fire in Canada's Boreal Forest: Paleoecology Results and General Circulation Model," *Canadian Journal of Forest Research* Vol 31, no. 5: 854–864.

21. Natural Resources Conservation Service (NRCS), "Land Degradation: An Overview," NRcS Soils. United States Department of Agriculture, 2001. Accessed online June 23, 2011.

22. Bartley Kives, "State of Emergency Declared in Winnipeg," *Ottawa Citizen* Apr 17, 2009: A5; Larry Kusch, Bartley Kives, "2009 Flood Compared to 1997 Flood," *Vancouver Sun* Apr 18, 2009: B5; "Manitoba Flood Update," *Winnipeg Free Press* May 12, 2009: A5.

23. Linda Nguyen, "Spring Flooding Crosses Canada," *Leader Post (Regina)* Mar 27, 2008: A11.

24. Jason Madger, "Conditions Too Dangerous for Divers to Search for Missing Gaspe Woman," *The Gazette (Montreal)* Aug 11, 2007: A8.

25. Jason Madger, "Flood Victims Want to be Relocated," *The Gazette* Aug 12, 2007: A3; Dominique Blain, "Red Cross Makes Pitch for Flood Victims," *The Gazette* Aug 22, 2007: A9.

26. *Climate Change Impacts and Adaptation*: 115.

27. "Maritimes Survey Damage of Storm," *Prince George Citizen* Jan 24, 2000: 6.

28. "Storm Hammers Maritimes Again," *Prince George Citizen* Jan 22, 2000: 7.

29. Cited in Adam Sweeting, *Beneath the Second Sun: A Cultural History of Indian Summer.* (Hanover: University of New Hampshire, 2003): 97.

30. Karen Smoyer-Tomic et al., "Heat Wave Hazards: An Overview of Heat Wave Impacts in Canada," *Natural Hazards* 28 (2003): 463.

31. "Ontario Breaks Power Consumption Record: Heat Wave Likely to Last for Weeks," CBC News, July 13, 2005. Accessed May 23, 2009 at ww.cbc.ca/canada.

32. Franz Boas, Ludger Muller-Wille, *Franz Boas among the Inuit of Baffin Island, 1883–1884: Journals and Letters.* (Toronto: University of Toronto Press, 1998): 54.

33. David Ljunggren, "Record Heat Forces Closure of Canada Arctic Park," Reuters, Aug 1, 2008. Accessed May 23, 2009 at alertnet .org.

34. A lengthy translation of Erman's account appears in S. W. Muller, *Frozen in Time: Permafrost and Engineering Problems.* (Reston, VA: American Society of Civil Engineers, 2008 [original edition 1947]): 6.

35. Ed Struzik, *The Big Thaw: Travels in the Melting North.* (Mississauga, ON: John Wiley, 2009): 245.
36. John Muir, *Writings of John Muir: Travels in Alaska.* Vol 3, M. Parson et al., eds. (New York: Houghton Mifflin, 1917): 40–41.
37. V. W. Bladen, *Canadian Population and Northern Colonization.* (Toronto: University of Toronto Press [for the Royal Society of Canada], 1962): 72.
38. Kirk Stone, *Northern Finland's Post-War Colonizing and Emigration.* (The Hague: Martinus Nijhoff, 1973): 49.
39. Balzhan Znimbiev, *History of the Urbanization of a Siberian City.* (Cambridge: Cambridge University Press, 2000): 88; Henry Morton, Robert Stuart, *The Contemporary Soviet City.* (Armonk, NY: M. E. Sharpe, 1984): 122.
40. Claus-M. Naske, Herman Slotnick, *Alaska: A History of the 49th State.* (Tulsa: University of Oklahoma Press, 1994): 112–14. See also Kirk Stone, *Alaskan Group Settlement: The Matanuska Valley Colony.* (U.S. Dept. of the Interior, Bureau of Land Management, 1950): 1–95.
41. Naske and Slotnick, *Alaska: A History:* 115.
42. Paul R. Ehrlich, *The Machinery of Nature.* (New York: Simon and Schuster, 1986): 56–57.
43. See Carl Zimmer's excellent article of January 23, 2007 in the *New York Times* for a really useful overview of "assisted migration." Retrieved online on May 27, 2009 from nytimes.com. The article Mr. Zimmer refers to that frames the debate over assisted migration is by Jason McLachlan et al., "A Framework for Debate of Assisted Migration in an Era of Climate Change," *Conservation Biology* 21 (2007): 297–302. But see also Anthony D. Barnosky, *Heatstroke: Nature in an Age of Global Warming.* (Washington, DC: Island Press, 2009): 199–200 for a highly readable account.
44. Aldo Chircop, "The Emergence of China as a Polar Capable State," *Canadian Naval Review* Vol 7 (1) Spring 2011: 1–14. Accessed on May 4 from navalreview.ca.

Conclusion

1. Lord Richard Skidelsky, Edward Skidelsky, *How Much is Enough? The Love of Money and the Case for the Good Life.* (London: Penguin: 2012): 131.
2. Ibid., 127.

3. Ibid., 183.

4. Sources of statistics used throughout this paragraph follow: 44% of Republicans believe in climate change: Pew Research Center, April 2, 2013. Downloaded on May 20, 2013 from people-press .org.; 40% of Republicans "worry" about climate change: Gallup Politics, "Republican Skepticism toward Global Warming Eases," April 9, 2009. Downloaded on May 20, 2013 from gallup.com; 60.9% of Americans believe in climate change: Pew Research Center, April 2, 2013. Downloaded on May 20, 2013 from people -press.org.

5. Yale Project on Climate Change Communication, "A National Survey of Republicans and Republican-Leaning Independents on Energy and Climate Change." Downloaded on May 20, 2013 from environment.yale.edu.

6. Texas Climate News, "New Polls: Most Accept Scientific View That Earth's Climate Is Changing," April 17, 2013. Downloaded May 20, 2013 from texasclimatenews.org.

7. Jane George, "Robins Roost in Kuujjuaq as Climate Changes: Bird on a Wire Says It's Spring," *Nunatsiak News* June 4, 2009. Downloaded May 20, 2013 from nunatsiaqonline.ca/.

8. Suzanne Goldenberg, "America's First Climate Refugees: Can a Baked Alaska Deny Climate Change?" *Grist* May 17, 2013. Downloaded on May 19 from grist.org.

9. CBC News, "Tuktoyaktuk on Front Line of Climate Change," Sept 9, 2009. Downloaded May 19, 2013 from cbc.ca/.

10. Skidelsky and Skidelsky, *How Much Is Enough?*: 124.

11. Thomas C. Peterson et al., eds., "Explaining Extreme Events of 2100 from a Climate Perspective," *Bulletin of the American Meteorological Society* (BAMS) July 2012: 1053.

12. Joseph. A. Tainter, *The Collapse of Complex Societies*. (Cambridge: Cambridge University Press, 1988): 1.

13. Charles Robert Darwin, *Life and Letters*. (London: D. Appleton & Co, 1888): 282.

Index

A

acceptance of climate change, 1–2, 139, 149, 168–169, 170, 220
African Americans, 98–102, 145–146
The Age of Consequences, 28–29
agriculture
    California, 15–16, 26–27, 60–61
    climate change impact, 72, 76
    farm workers, 13–15
    irrigation, 62, 69–71
    Mexico, 36–39, 49–51
    migration of workers, 15–16
    Southwest and Great Plains, 8–9, 10–15, 70–71
    technology and mechanization, 10–12, 13–15, 69–70
    water scarcity, 69–71, 76
air conditioning, 113, 117, 154–155
Alaska, 214–217, 222
Alberta, 199–202
Alvarez, Daniel, 135
Anasazi, 3–4
Andrews, Jim, 154
Angell, Charles, 11
Angus, Julie, 109
anthropogenic cause of global warming, 140, 158, 220–221
aquifers, 63–64, 67–68
Archer, Christine, 172–173
Arctic ice, 89–90, 107
Arctic region of Canada, 208–211, 217–218
Arizona, 32–33, 52, 54–55
Arrhenius, Svante, 139–140
assisted migration, 213–214, 217–218
Astrodome (Houston), 97–98, 99

Atlantic Provinces, 206
Auyuittuq National Park, 208

B

Banda, Victor Hugo Rascón, 44
Bangladesh, 104
Barry, John, 100
bay checkerspot butterfly, 217
Bennett, H.H., 10–11
Bhola cyclone, 104
blacks, 98–102, 145–146
Bloomfield, Janine, 110–111
Bluestein, Howard, 174–175
Boas, Franz, 208
boll weevil, 145–146
Booth, William, 46–47
border between United States and Mexico, 31–32
boreal forest, 203
British Columbia, 211–212
Brown, Lester R., 18
Buffalo Commons, 72, 74–75
bum blockade, 22–23
Bush, President, 102

C

Caldeira, Ken, 172–173
California
    agriculture, 15–16, 26–27, 60–61
    cryosphere, 76, 77
    economic collapse, 57–59
    illegal immigration, 45–46
    legislation and politics, 27
    megafires, 82–84
    Mexican migrants and workers, 20, 27, 42, 45–46

Okies migrants, 15–16, 21–27
outmigration, 57–58, 83–84, 192–193
treatment of migrants, 22–26
water problems, 57, 59–60, 76–82
Canada
  Alberta and Saskatchewan, 199–
    202
  Arctic region, 208–211
  British Columbia, 211–212
  deglaciation, 200
  drought, 199–202
  effects of climate change, 194–195,
    199–200, 202, 204–205, 208,
    210
  flooding, 204–205, 226–227
  freshwater reserves, 191–192
  heat waves, 207–210
  Manitoba, 204–205
  Maritime Provinces, 206
  megafires, 202–204
  migration from United States,
    xiv, 163, 192–194, 195, 196–199,
    210–213
  Northwest Territories, 210–211
  Quebec, 202–203, 205
  reaction to US migrants, 198
  relocation of communities, 222
  sea level rise, 205–207
  southern prairie region, 199–200
capitalism, 8–9
carbon dioxide (CO_2), 140
Category 5 storms, 92, 93–94
chain-wells, 62, 68, 238
Chavez, Leo, 54
Cherokee, 7–8
Chew, Sing C., 18, 48–49, 223
Chicago, 1995 heat wave, 129–132,
  134–138
cholera infantum, 123
Ciudad Juárez, 46–47
Climate Central, 181–182
climate change
  acceptance, 1–2, 139, 149, 168–169,
    170, 220

agriculture and, 72, 76
in Arctic region, 208–210
aridity of Southwest, 72
in Canada, 194–195, 199–200, 202,
  204–205, 208, 210
denial, 142–143
disasters frequency, 109–110
economy and, xii–xiii, 33, 225–226,
  227
evidence, 28, 141–142, 225–226
extreme storms and sea level rise,
  88–89, 92–93
extreme weather, 157–161
heat waves, 114–115, 136–138, 161–
  162
hurricanes, 92–93
impacts, 28–29, 72
methane, 141
migration as response, 220–222
preparation and adaptation, 143–
  144
refugees, 107
scientists and, 143, 157–161, 178, 225
speed of changes, 189–190
Superstorm Sandy, 186–187
technology for mitigation, 224–225
tornadoes, 172–173
water reserves, 76
climate migration
  beginnings in United States, 2–4
  evidence, 221–224
  overview and forecast, xii–xv
  as response to climate change,
    220–221
climate refugees, 18–19, 107, 222–223
CO_2 (carbon dioxide), 140
collapse in North America, 180, 227
Colley, Jack, 97
colonies for migration, 213–218
Colorado, 76
Colorado River, 78–79
cotton industry, 146
cryosphere, 67, 76, 77, 78–79
Cuomo, Governor, 160, 166, 181

D

Daley, Richard M., 130–132, 134–136, 137

Darwin, Charles, 227

Davis, James "Two Gun," 22–23

De Gaetano, Art, 186

deglaciation, 89–90, 107–108, 185, 200

denial of climate change, 142–143

desalination of water, 79–80

Detroit, 179–180

Diamond, Jared, 48

disaster preparation
Bangladesh, 104
climate change and, 143–144
evacuation, 105–106
heat waves, 132, 134–135, 137–138
lack of, 136
New York area, 110–111
United States, 102–104, 105

disasters, migration as response, 145, 149–150, 162–163, 220–222

Donoghue, Edmund, 130–132, 135–136

Douglas, Paul, 133

drought
Canada, 199–202
consequences and examples, 155–157
description, 153
dryness and, 152
Mexico and United States, xi–xii, 40
permanence, 155
refugees, 16–17
severity and outlook, 162
survival strategies, 73–74
Texas in 2005–2011, 148–151, 161
trees killed, 150–151

drownings during heat waves, 128–129

Dust Bowl, 5, 10–11, 13–15, 222

dust storms, Southwest and Great Plains, 4, 5–8, 12–13, 17

E

East Coast, sea level rise, 185

Eckels, Robert, 97–98, 99

economy
collapse, xii–xiii, 18, 33, 48–49, 57–59
drought and, 156–157
environmental challenges, 83
extreme weather events, 161–162, 181, 225–226, 227
illegal workers, 52
Mexico and Mexican migration, 32–36, 39, 47–48
reason for migration, 146–147

EF5 tornadoes, 166–167, 177, 226

electrical grid, 113, 131, 155

electric fan, 126–127

El Niño, 64–65

employment, illegal workers, 51–52

environmental refugees, 18–19, 107, 222–223

Erman, Georg, 209

Europe, 2003 heat wave, 130, 161

evangelism, 168–169, 177–178

evapotranspiration, 152

explorers in North America, 190–192

extreme storms
Category 5, 92, 93–94
climate change and, 88–89, 92–93
frequency, 109–110
prediction, 94–95
South Atlantic, 93–94
strength and timing, 165

extreme weather
climate change, 157–161
evidence of changes, 141–142
frequency, 159–160, 162
impact on economy, 161–162, 181, 225–226, 227
migration as response, 162–163

F

farm workers, 13–15

fence between United States and Mexico, 31–32

Finkelstein, Max, 202–203

Finland, 213–214

Finley, John Park, 168, 175
Fischetti, Mark, 94
flooding, 107, 204–205, 226–227. *see
    also* storm surges
Ford company, 11–12
forest fires. *see* megafires
freshwater. *see* water

G
Gaspé (Quebec), 205
generosity of people, 4–5, 21–22,
    98–99
Gibson, Jabar, 96–100
Gleick, Peter, 183
global warming. *see* climate change
Goldenberg, Suzanne, 221–222
Greater New York. *see* New York
Great Hurricane of 1938, 108–109
Great Lakes, 195
Great Migration, 145–146
Great Plains
    agriculture, 8–9, 10–15, 70–71
    aquifers, 63–64
    dust storms and Dust Bowl, 4, 5,
        7–8, 12–13
    importance, 66
    outmigration, 15–17, 19–22,
        222–223
    population decline, 66–67, 71–72
    tornadoes, 171, 174
    water and irrigation, 66–72
"The Great Plains: From Dust to
    Dust" (F. and D. Popper), 66
greenhouse effect, 139–141
greenhouses, 139–140, 249
Greenland, 107–108, 185
Green Revolution, 36–37, 38
Guatemalans in United States, 41

H
haboobs, 154. *see also* tornadoes
Hayden, Mike, 73
Hayhoe, Katharine, 183
heat, 153–154, 174, 175–176, 182–183

heat-island effect, 119–121
heat stroke. *see* hyperthermia
heat waves
    in Arctic region, 208–209
    Canada, 207–210
    Chicago in 1995, 129–132, 134–136,
        137
    climate change and, 114–115, 136–
        138, 161–162
    consequences, 113–114
    cooling in, 118–119, 126–129
    deaths from (*see* hyperthermia)
    disaster preparation, 132, 134–135,
        137–138
    Europe in 2003, 130, 161
    fires and, 114
    frequency and impact, 114–115,
        124–125, 224
    La Niña and, 125
    New York in 1896, 115–119, 121–124
    of 1936, 126–127
    technology, 125–126, 127
Hoeme, Fred, 8
horses deaths in heat waves, 117–118,
    121–122
Hundley, Norris, 59–60
hurricanes
    climate change and, 92–93
    heat and, 182–183
    New York in 1938, 108–109
    power of, 183–184, 186–187
    risks, 94–95
    sea surface temperatures, 93–94,
        183
hyperthermia
    Chicago heat wave, 130–132, 135–
        136
    criteria, 130–131
    diagnosis, 132–133
    effect on body, 128, 129
    ignorance about, 114, 115, 132–134
    New York in 1896, 122–124
    predictions, 138
    treatment, 127–128

I

Ibom, 177, 255
ice-melt, 89–90, 107–108, 185, 200
illegal immigration, 45–46, 51–52
insurance companies, 161–162, 180–
    181, 225
internal migration within United
    States, 26–27, 29
IPCC (Intergovernmental Panel on
    Climate Change), 89, 162–163
irrigation, 62, 69–71

J

Jackson, Reverend Jesse, 101–102
Jarraud, Michel, 155
jet stream, 172–173
Johnson City (Kansas), 43–44

K

kangaroo rats, 84–85
Katrina
    description, 92, 93
    evacuation of New Orleans, 96–97,
        105–106
    impact on United States, 91–92
    treatment of victims, 98–100
Kerry, John, 172
Klinenberg, Eric, 129, 136
Kotlikoff, Laurence, 147

L

"la crisis," 33–36, 223
La Niña, 64–65, 125, 175
Latin Americans in United States,
    xiii, 54–56
    migration to, 41, 45, 56
    push and pull factors, 33–34, 39
LeDuff, Charlie, 179–180
Legge, Alexander, 12
Leopold, Aldo, 5, 8, 178
Lerner, Michael, 25–26
lifeboat ethics, 22, 25, 98–99. *see also*
    generosity of people
Lindsay, Dwight and Marg, 151–152

Los Angeles, 76–78, 81
Lynas, Mark, 111, 138

M

Malin, James, 7
Manitoba, 204–205
man-made global warming, 140, 158,
    220–221
Maritime Provinces, 206
Matanuska (Alaska) settlers, 215–217
McLemen, Robert, 198–199
McManus, Gary, 3
McQuaid, John, 94–96
megafires
    California, 82–84
    Canada, 202–204
    heat waves and, 114
    human cost, 151–152
    Texas, 151
    trees and, 151
    United States, 82–83
Melville, Roberto, 49
methane, 141
Mexican migration
    to California, 27, 42, 45–46
    for economic salvation, 32–33, 47–48
    illegal immigration, 45–46, 51–52
    push and pull factors, 33–34, 39–
        40, 47, 53–54
    repatriation in 1930s, 20
    to United States, 31–33, 40–41, 42–
        43, 51–54, 223
Mexicans in United States
    as force, 54–55
    illegal workers, 51–52
    population, 32, 41, 45–46
Mexico
    agriculture and hunger, 36–39, 49–51
    climate change and drought, xi–
        xii, 40
    collapse, 227
    economic problems, 34–36, 39
    environmental devastation, 36–38,
        40, 49–51, 53, 223

"la crisis", 33–36, 223
　view of United States, 47–48
Middenhorf, Alexander Theodor von, 209
migrant communities, 214–217
migrants
　first to migrate, 20–21, 146–148, 197
　impact on new place, 28–29
migration as response to disaster, 145, 149–150, 162–163. *see also* outmigration
Mississippi River, 156–157
　flood of 1927, 101–102
Morse, Charles, 119
Moscow heat wave, 113–114
Muir, John, 212
Muller, Siemon W., 208–209
Mumme, Stephen, 49
Munich Re, 161, 225

N
New Orleans
　disaster preparation, 105
　impact of Katrina, 91–92
　Mississippi flood of 1927, 101–102
　storm safety and evacuation, 96–99, 105–106
　treatment of victims, 98–102
　vulnerability, 94–96
New York
　disaster preparation, 110–111
　fires and heat waves, 114
　Great Hurricane of 1938, 108–109
　heat wave of 1896, 115–119, 121–124
　impact of extreme weather, 181–182
　outmigration, 182, 196
　sea level rise and flooding, 91, 108, 110–111, 186
　Superstorm Sandy, 181, 182–184, 186–187

North, Canadian, 208–211, 217–218
North Atlantic, storms in, 109
Northwest Territories, 210–211
Noss, Reed, 75

O
Ogalla Aquifer, 66–68, 71–72, 222–223
Okies, 21–27
Olson, Culbert, 27
one-way disk plow, 11
Orlov, Dmitry, 180
outmigration
　California, 57–58, 83–84, 192–193
　Great Plains and Southwest, 15–17, 19–22, 222–223
　New York, 182, 196
　as response to disaster, 145, 149–150, 162–163

P
Pacheco, José Emilio, 33
Pearman, Graeme, 107
permafrost, 208–209
Peters, Scott, 81
Pine Island Glacier, 89–90
Plain Indians and tornadoes, 7–8, 177, 255–256
plows, 11
Popper, Frank and Deborah
　on population decline, 66–67, 71–73, 74
　reaction to, 72–73, 74
　reintroduction of species, 66, 72, 74–75
Portes, Alejandro, 47
Powell, John Wesley, 69–70
power plants, 113, 131, 155
preparation. *see* disaster preparation
Prince George (British Columbia), 212
prosperity and migration, 146–147
pumping of water, 69–71

Q

qanats, 62, 68, 238
Quebec, 202–203, 205
Quintero, Rita, 43–44

R

racial politics and racism, 4–5, 25,
 98–102
radiation stress, 185–186
Reed, Candice, 58
refugees
 attitude towards, 102–103
 from climate, 18–19, 107, 222–223
 from Dust Bowl, 16–17
 use of term, 16–18, 101–102
reintroduction of species, 66, 72,
 74–75
relocated settlers, 214–217
relocation of communities, 222
Republicans, 56
Resettlement Administration, 214–
 215
rewilding, 66, 72, 74–75
Romney, Mitt, 143
Roosevelt, Franklin, 214
Roosevelt, Theodore, 119, 123
Rossy, Fran, 205
Rub' al-Khali, 6–7
Russia, 113–114, 213–214, 218

S

Saffir-Simpson scale (SS), 92, 184
Sallenger, Asbury, 111–112
Samaras, Tim, 171, 173
Sanders, Jerry, 81
San Diego, 79–80, 81
Sandy, 181, 182–184, 186–187, 226–227
Saskatchewan, 199–202
Saudi Arabia, 5–7, 80
SB 1070 (Arizona), 32, 52–53
Schleifstein, Diane, 106
Schleifstein, Mark, 94–96, 105–106
Schrag, Peter, 58–59

scientists, on link between climate
 change and extreme weather, 143,
 157–161, 178, 225
sea level rise (SLR)
 Canada, 205–207
 deglaciation and, 89–90, 107–108,
 185
 displacement of people, 107
 forces involved, 111–112
 impact, 108–109, 183
 New York area and East Coast,
 110–111, 185
 outmigration and, 196
 predictions, 110, 195–196
 temperatures and, 88–89
 vulnerable areas, 90–91, 108–109
Sears, Paul, 10
sea surface temperatures, 93–94,
 182–183
semi-arid regions, 63–64, 84–85
Sensenbrenner, Jim, and legislation,
 27, 54
Shergin, Fyodor, 209
Simon, Joel, 38
Sioux, 7, 177
Skidelsky, Richard and Edward, 219–
 220, 224
Smokey the Bear Effect, 151
Smythe, William E., 62–63
social capital, 20–21, 62–63
Soulé, Michael, 75
South Atlantic, extreme storms and
 sea level rise, 93–94
Southwest
 agriculture, 8–9, 10–13, 70–71
 climate change and aridity, 72
 dust storms, 7, 12–13
 La Niña and El Niño, 64–65
 natural adaptations, 73–74, 84–85
 outmigration, 15–17, 19–22, 222–223
Steinbeck, John, 26
Stern, Sir Nicholas, 197
Stone, James, 203

storms. *see* extreme storms
storm surges, 183–184, 185–186, 206, 226–227
Struzik, Ed, 210
supercell thunderstorms, 174–175, 176–177
Superdome (New Orleans), 97
Superstorm Sandy, 181, 182–184, 186–187, 226–227
Super Tuesday tornadoes, 169–172
Suzuki, David, 107

T
Tainter, Joseph, 227
Tarahumara people, 43
Taylor, Paul, 15–16
technology
  in agriculture, 10–12, 13–15, 69–70
  air conditioning, 113
  climate change mitigation, 224–225
  in heat waves, 125–126, 127
  progress and impacts, 9–10, 13–15, 28, 38, 70–71
  pumping of underground water, 69–70
temperatures, 88–89, 93–94, 175–176. *see also* heat waves
  at sea surface, 93–94, 182–183
Texas, 115, 148–151, 161
Thunder Being, 7–8, 13
Times-Picayune series "Washing Away," 94–96, 105
tornadoes
  climate change and, 172–173
  deaths and damage, 154, 171, 177
  EF5, 166–167, 177, 226
  formation, 173–174, 176–177
  impact, 167–168
  intensification, 154–155, 166–167, 177, 225–226
  Plain Indians, 7–8, 177, 255–256
  on Super Tuesday, 169–172
  in winter, 170, 172

tractors, 11–12
transhumance, 43
trees, in drought and megafires, 150–151
Trenberth, Kevin, 63, 110, 186–187
TriState. *see* New York
Tugwell, Rex, 214, 215

V
Vancouver (British Columbia), 211
van Heerden, Ivor, 103
Van Tilbrug Clark, Walter, 153
växthaus, 139–140, 249

W
"Washing Away" series, 94–96, 105
wastewater recycling, 80–82
water. *see also* drought
  availability on Earth, 61–62
  California, 57, 59–60, 76–82
  Canada, 191–192
  climate change and, 76
  cryosphere, 67, 76, 77, 78–79
  desalination, 79–80
  for electricity generation, 154–155
  Great Plains, 66–72
  importance, 61, 62–63
  La Niña and El Niño, 64–65
  recycling, 80–81
  reserves, 67–69
  scarcity and disappearance, 63, 69–71, 76, 192–193
Weeks, Gregory and John, 51–52
Westhoven, William, 184
Whiteford, Scott, 49
wildfires. *see* megafires
Williams, Reuben, 36
wind erosion, 5, 8
Wolff, Alfred, 117
Worster, Donald, 8–9

Y
young Americans and families, as migrants, 20, 146–148, 197

About the Author

As he turns 60, Giles Slade is still busy learning volumes from his mistakes. He has three fine sons and lives on the West Coast of Canada. Sandra and Giles have been "an item" for 33 years this fall, and someday they will sail to Hawaii on a small but affordable boat called the Follow-Your-Heart. *American Exodus* is his third book. There will be more.

If you have enjoyed *American Exodus*, you might also enjoy other

Books to Build a New Society

Our books provide positive solutions for people who
want to make a difference. We specialize in:

Sustainable Living ✦ Green Building ✦ Peak Oil
Renewable Energy ✦ Environment & Economy
Natural Building & Appropriate Technology
Progressive Leadership ✦ Resistance and Community
Educational & Parenting Resources

New Society Publishers
ENVIRONMENTAL BENEFITS STATEMENT

New Society Publishers has chosen to produce this book on recycled paper made
with 100% post consumer waste, processed chlorine free, and old growth free.

For every 5,000 books printed, New Society saves the following resources:[1]

27	Trees
2,447	Pounds of Solid Waste
2,692	Gallons of Water
3,512	Kilowatt Hours of Electricity
4,448	Pounds of Greenhouse Gases
19	Pounds of HAPs, VOCs, and AOX Combined
7	Cubic Yards of Landfill Space

[1]Environmental benefits are calculated based on research done by the Environmental Defense Fund and
other members of the Paper Task Force who study the environmental impacts of the paper industry.

For a full list of NSP's titles, please call 1-800-567-6772 *or check out our web site at:*

www.newsociety.com